Absolute Climate Change

by Rolf A. F. Witzsche

Text Copyright (c) 2015 - Rolf A. F. Witzsche

Contents

3

About the Illustrated Science series:
Ice Age – Climate Change
and the book
Absolute Climate Change

Numerous fields of evidence tell us that the next Ice Age is near. Most of the evidence was discovered in the 1990s and thereafter. Some evidence is measured in ice cores; some is measured in space, by satellites. Some measurements are also made on the ground in terms of measurements of the Earth's magnetic-pole drift observed in northern Canada. All of this is seen combined with high-energy physics experiments at a leading national laboratory, and is also explored in the small in static experiments.

Against the background of these widely diverse types of evidence that have been recently discovered, the historic Little Ice Age in the 1600s, takes on a new dimension as a yardstick for measuring the future that by this evidence promises to be up to 40-times colder than the Little Ice Age had been. It qualifies for the term, Absolute! The evidence poses a great challenge ahead. Are we ready to respond? The Ice Age phase shift in climate is a stark in differences as night and day, and similarly fast.

In the Little Ice Age between 10% and up to 30% of the populations in Europe had perished by starvation. The last Big Ice Age was evidently vastly harsher. Only 1-10 million people emerged from it alive. That's all we had after 2 million years of development. We want to do far better this time around; and we can, with large-scale technological infrastructures for our food supply. But will we create them? Will we get the job done in the 30 years that we still have left before the Ice Age starts anew? Will we even consider it? And how certain are we that the phase shift to the next glaciation period will begin, as the evidence suggests, in the 2050s? We have no slack on this front. By failing itself on this absolute front, humanity actively commits suicide.

So, what will the answer be? Will we move with the evidence? Or will we lay ourselves down to die by default?

It takes an independent researcher to brake the taboos that have kept mainstream cosmology imprisoned, increasingly, during the past century, even while what is regarded as taboo is known to be wrong.

The Illustrated Science series is intended to open the scene beyond the threshold of accepted taboos, to where the actual physical evidence speaks for itself.

The scope of the existential challenge that the Ice Age brings with it, takes astrophysics out of the academic domain and places it into the foreground as one of the most-critical issues of our time. The big Climate Change events that have already worldwide effects are mere fringe effects in the flow of the ever-changing cosmic dynamics. The big effect, when the Ice Age begins anew, promises to be caused by a dimmer and colder Sun with 70% less radiated energy. This defines our climate future.

Sure, we can live with all that by creating new platforms for agriculture that are able to operate under Ice Age conditions. But will we do it? The task is enormous. Or will we fail ourselves on this front? We have no reason to allow us to fail. We have the materials and energy resources on hand to accomplish everything that is required for us to continue to live in an Ice Age World. But will we do it? The big question that never goes away, therefore, is; will we develop our inner resources as human beings sufficiently to get the job done, and to get it done in time? Or will we do nothing, ignore the challenge, and condemn our children and one-another to an agonizing death by starvation? That's the choice.

Towards meeting the inner challenge, I have created the epic series of novels, The Lodging for the Rose. And further, towards meeting the science challenge, I have produced numerous research books and several dozen exploration videos that the Illustrated Science series is modeled after. The work is the result of a quarter century of research, for which numerous elements of evidence in related fields came to light during the timeframe of my research.

It is my hope that the work that went into all of these projects will help in some degree - for humanity that we are all a part of - to write itself a ticket to have a future.

High-resolution color images, of the images in this book, can be obtained at www.iceagetheatre.ca

*Something changes in the world

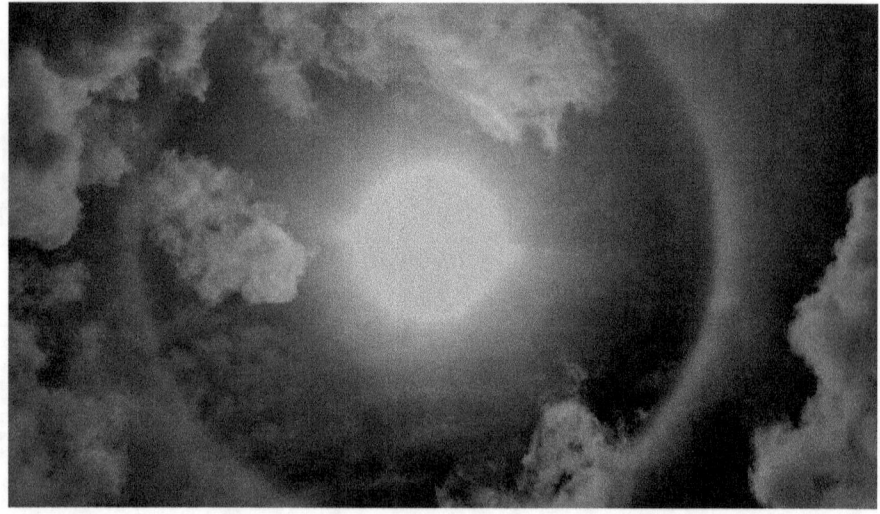

Every once in a while something changes in the world that changes everything, so it seems, that changes even history. A profound discovery of truth has this potential.

A spark of fundamental truth

A spark of fundamental truth, that when it overturns a concept that has been widely cherished for a long time, can invalidate a part of history that has never been true. It creates a new world that is radically advanced, so that the old world has no place in it. This video has the power to have such an effect.

What you see should not be possible

It begins with the image that you see before you. What you see should not be possible according to the most-widely accepted perception of the nature of our Sun, the solar system, the galaxy, and the universe. In fact, you yourself, should not exist.
The reason why I can say this with confidence, is rooted in something as basic as the nature of sunlight.

Scenes of vibrant colors

Scenes of vibrant colors bathed in richly white sunlight, as the one before you, are rather common all around the world. They have been seen by almost everyone.

If what is taught in the schools about the Sun is true

But if what is taught in the schools about the Sun is true, nothing what you see here should exist. Neither the image, or the object, nor the beholder itself, should exist. But they all do exist.

We face a paradox here

This means, we face a paradox here. The most fundamental scientific concept of the universe tells us that what we see here should not be possible. Our eyes tell us the opposite.

Three paradoxes:
The sunlight paradox
Paradox in climate change
A Science paradox

** Three paradoxes:

The sunlight paradox; the paradox in climate change; a Science paradox.

This video also explores two other related paradoxes. So, let's begin:

** The Sunlight Paradox

Sacredly held axioms in science, an impossibility

Just look at this image. As I said before, this simple scene of common flowers totally disproves the most widely accepted scientific concept about our Sun and the universe. That's astonishing. It renders one of the most sacredly held axioms in science, an impossibility, and suggests that as we wake up to the truth, we begin to realize that we haven't seen anything yet.

The old model of the Sun that is false

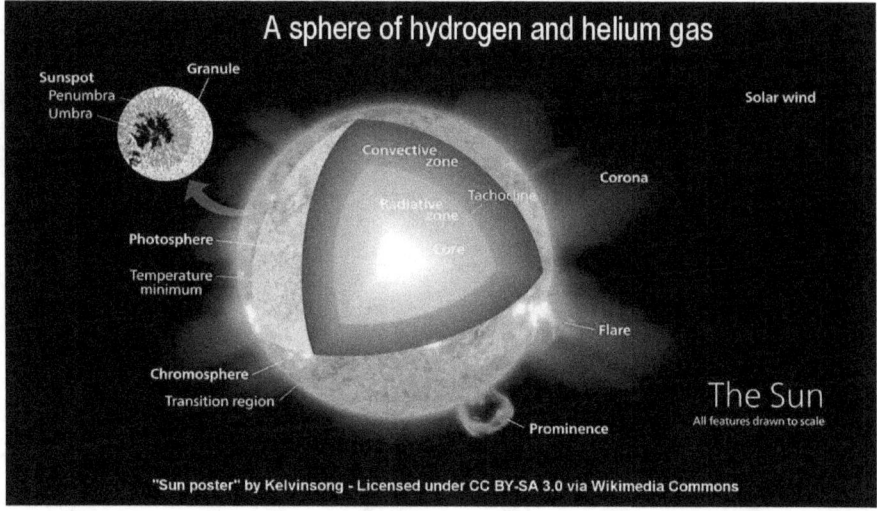

A sphere of hydrogen and helium gas

Sunspot
Penumbra
Umbra

Granule

Solar wind

Convective zone

Corona

Tachocline

Radiative zone

Photosphere

Core

Temperature minimum

Flare

Chromosphere

Transition region

The Sun
All features drawn to scale

Prominence

Let's consider the old model of the Sun that is false. It is almost universally accepted that the Sun is a sphere of hydrogen gas that is heated by nuclear fusion reactions occurring in its core that fuse hydrogen atoms into helium atoms, whereby heat is released that oozes to the surface on a 30 million years long path. That's the theory. It is taught in the schools. The model is printed on posters. But paper is patient. Anything can be printed on it. Schools can teach anything.

If the hydrogen-sun theory was true

You need to test the truth yourself. If the hydrogen-sun theory was true, would you be able to see what you see here? You wouldn't, because the hydrogen atom, no matter how hot it becomes, cannot produce the rich spectrum of light that makes the colorful scene possible.

Sunlight directed through a prism

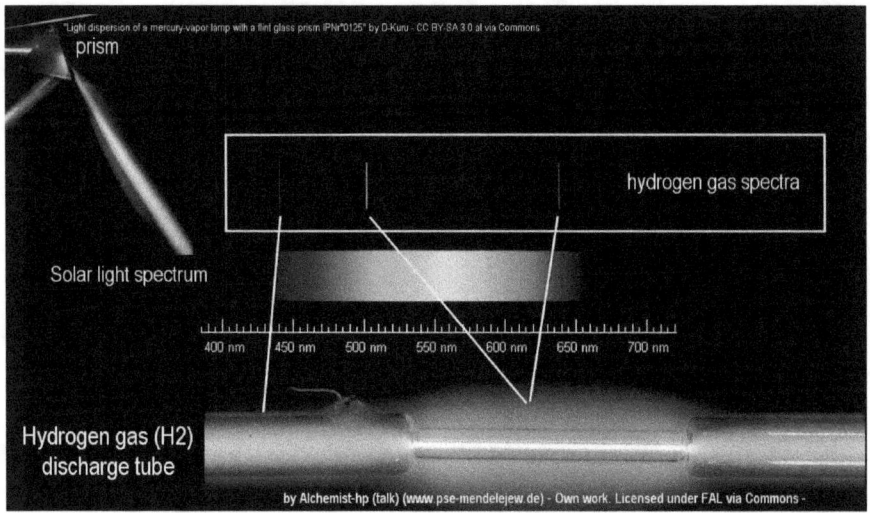

prism

hydrogen gas spectra

Solar light spectrum

400 nm   450 nm   500 nm   550 nm   600 nm   650 nm   700 nm

Hydrogen gas (H2) discharge tube

When the sunlight that illumines our world is directed through a prism, some basic physical principles come into play that split the white light into its constituent colors, nicely sorted out according to the wavelengths of the different colors that are combined in the light.

The simple experiment illustrates that the common sunlight contains within it all the visual colors in a wide continuous band. When we do the same with light emitted from hydrogen gas, we see a totally different spectrum. We see almost nothing. We see that hydrogen gas emits light in only 4 extremely narrow strips that cover only a few percent of the visible spectrum.

This rather meagre spectrum of light would be all that we would see emitted from the Sun if the commonly accepted perception of the Sun was correct, under which the Sun is a sphere of hydrogen gas. Obviously, since this is not what we actually see as sunlight, which our richly colorful world illustrates, it becomes self-evident by what we see, that the Sun that illumines our world simply cannot be a sun that is a hydrogen-gas sphere that would emit only a meagre light spectrum.

The entire green biosphere would not exist

It should be noted here that it would not only be the rich spectrum of colors that we would not see, if the old theory was correct, under which the Sun is a sphere of hydrogen gas heated from within. Most likely the entire green biosphere would not exist that depends on the services provided by the chlorophyll molecules that utilize sunlight to break down the atmospheric carbon gases for the nourishment of plants. Since all life depends on this process for its food, directly and indirectly, life would not exist if the widely accepted theory about the Sun was correct, nor would we exist without plants for our food supply.

Chlorophyll molecules absorb light only in two narrow regions

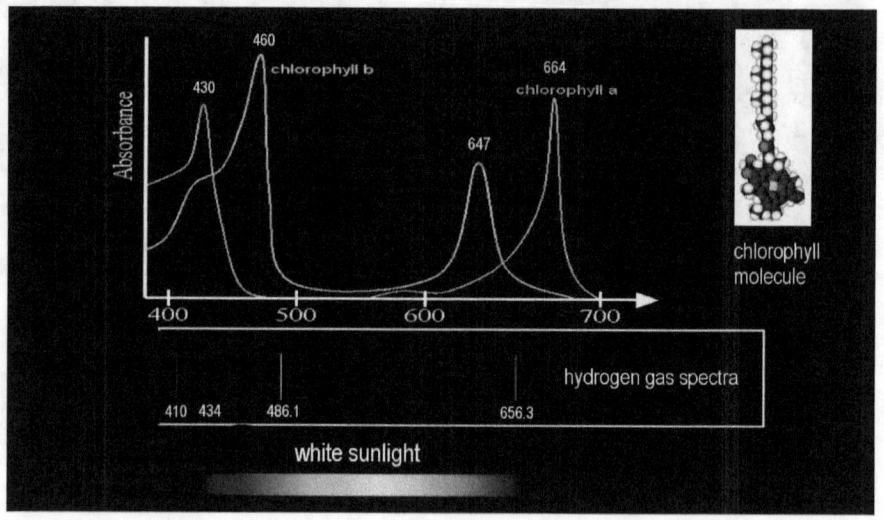

chlorophyll molecule

It has been discovered that the chlorophyll molecules do not utilize the full spectrum of the sunlight, but absorb light only in two narrow regions of the visible spectrum. It has further been recognized that the extremely narrow emission bands of the hydrogen light-spectrum fall outside the main chlorophyll absorption regions, so that the chlorophyll would not be activated by it. This means that the Earth would be a barren rock under a Sun that is a sphere of hydrogen gas, heated from within.

Accepted theory about the Sun is absolutely incorrect

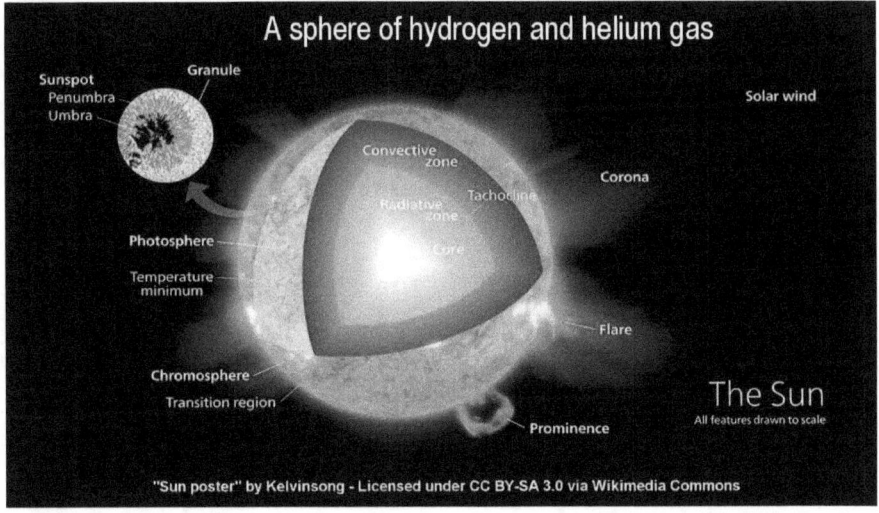

These rather striking facts, in basic comparison, prove that the entire accepted theory about the Sun is absolutely incorrect, because none of what we experience on the Earth is possible under a hydrogen sun, including life itself.

The very foundation of the theory of the internally powered Sun has thereby been rendered false. It requires the Sun to be a sphere of hydrogen gas that enables internal nuclear-fusion reactions to occur. All this stands disproved by the evidence that we see, even by life itself. Nothing but fairy-tale dust can bridge the fundamental contradiction.

Under the hydrogen-sun theory

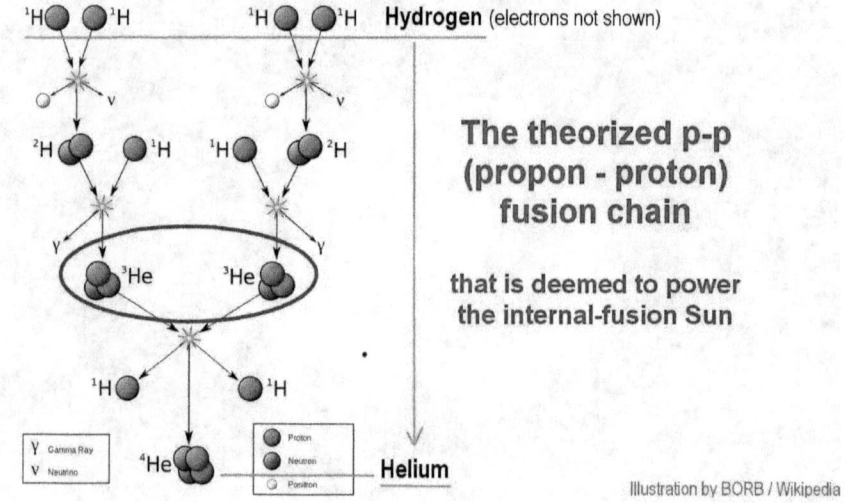

Hydrogen (electrons not shown)

The theorized p-p (propon - proton) fusion chain

that is deemed to power the internal-fusion Sun

Helium

Illustration by BORB / Wikipedia

Under the hydrogen-sun theory, the solar hydrogen atoms are fused into helium atoms at the core of a sun, with free energy being derived from the process that takes millions of years to ooze to the surface by slow convection, as the theory has it.

The fusion-chain concept has many flaws

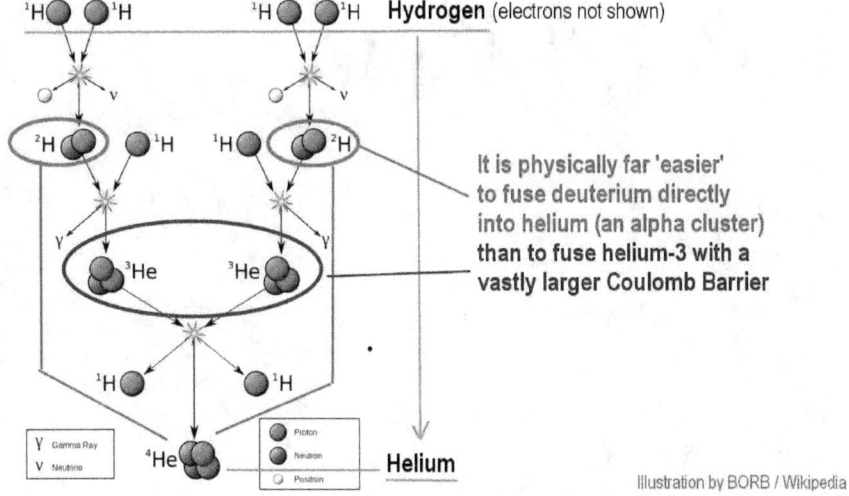

It is physically far 'easier' to fuse deuterium directly into helium (an alpha cluster) than to fuse helium-3 with a vastly larger Coulomb Barrier

Illustration by BORB / Wikipedia

While the fusion-chain concept that begins with hydrogen atoms has many flaws in itself, with numerous debatable scientific contradictions among them, the really big contradiction, is the absolutely black and white issue that by its evidence renders the basic premise of the theory impossible. The theory requires the Sun to be a sphere of hydrogen gas. All experienced evidence disproves this possibility.

29

Elegant as most people see it

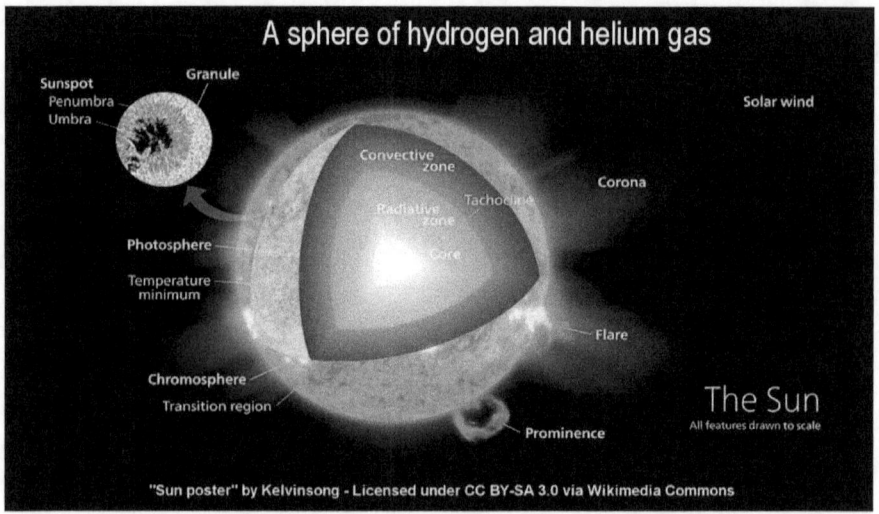

A sphere of hydrogen and helium gas

Sunspot
Penumbra
Umbra

Granule

Convective zone

Tachocline

Radiative zone

Core

Corona

Solar wind

Photosphere

Temperature minimum

Chromosphere

Transition region

Flare

Prominence

The Sun

All features drawn to scale

"Sun poster" by Kelvinsong - Licensed under CC BY-SA 3.0 via Wikimedia Commons

The accepted theory of the internally heated nuclear-fusion-sun is rendered absolutely false by the simple fact that the white sunlight that we see every day, would not be possible, and by the fact that we live, which should not be possible either.

This means that the most-widely accepted theory in cosmology, has the foundation that it stands on disproved by something as common in nature as the sunlight. The tragically accepted theory about the Sun, which is still being accepted - elegant as most people see it - stands absolutely rendered false, a deeply fundamental paradox that simply cannot be resolved.

The full spectrum of the sunlight

The full spectrum of the sunlight that we see all around us, cannot by any process that we know, originate from a hydrogen-gas Sun. Likewise can't the green chlorophylls in plants operate in hydrogen-light that lacks the wide spectrum to activate it. Nothing can bridge the paradoxes that the old, flat-earth-type perception of the Sun sets up.

Why the hydrogen-gas spectrum is as meagre

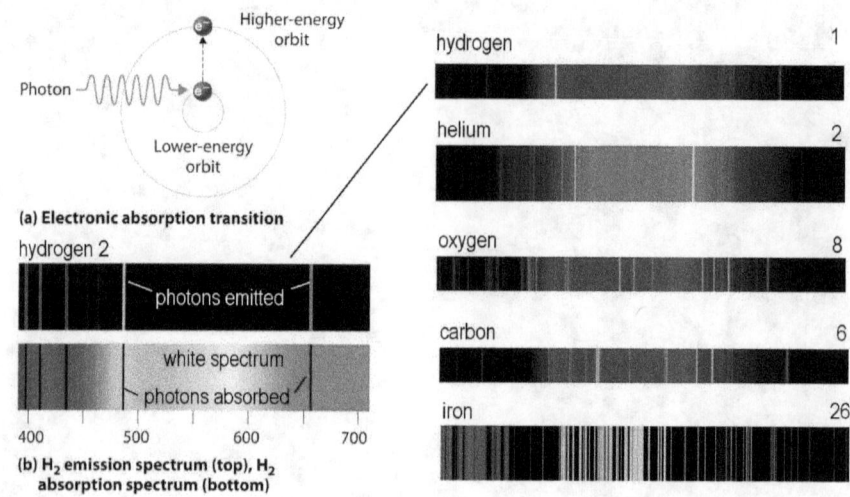

(a) Electronic absorption transition

(b) H₂ emission spectrum (top), H₂ absorption spectrum (bottom)

The physical reason why the hydrogen-gas spectrum is as meagre as it is, and far too meagre to support life, is rather simple. When the single electron of a hydrogen atom absorbs energy, either from a compatible photon, or from electric interaction with flowing plasma, the energy that is absorbed by the electron lifts the electron to a higher orbit that corresponds with a higher-energy state. Since the resulting higher orbit is unnatural for the hydrogen atom, the electron snaps back to its original orbit and in the process emits the stored up excess energy in the form of a photon, which is the carrier of light.

The excitation of electrons in an atomic structure is the common process by which light is produced in the universe. With the hydrogen atom having only one single electron in its surrounding orbital shell, the possibilities for its excitation are rather few. When larger atoms, that are made up of numerous electrons, in numerous shells, and sub-shells, are subjected to external excitation, the range of the possibilities for such excitation widens dramatically, both for the electrons to capture energy and to emit photons of different colors of light. The type of difference is dramatically evident in the case of iron. The iron atom has 26 electrons occupying three shells with numerous sub-shells. An atom of such a complex structure, evidently can emit a large number of different colors of light. But here too, the light that is emitted by it is broken up into isolated strips with voids and gaps between them. All this means that for the seamless band of all the colors that constitutes white light, as we get it from the Sun, a large variety of different atomic elements must be present in the solar photosphere where the sunlight is produced. The old model of the Sun cannot accommodate this fundamental requirement for white sunlight to be possible.

Note 1: The image on the left is based on http://chemwiki.ucdavis.edu/@api/deki/files/27812/b9bbdd38b91f87f39070da7115830b86.jpg?

Note 2: Some of the emission spectral lines on the right (from Wikipedia) have the full-color spectrum as a background for reference purposes.

The emission-spectra of helium and hydrogen together

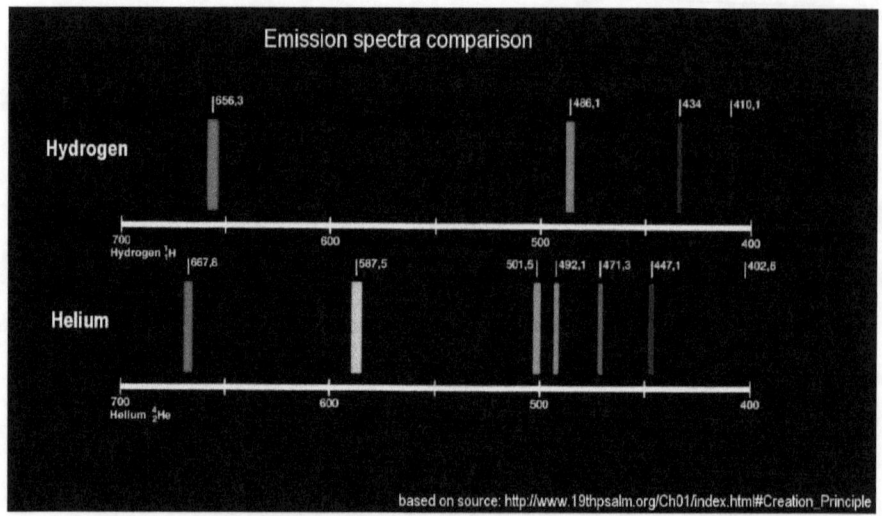

Even if one adds the emission-spectra of helium and hydrogen together, since 25% of the Sun is theorized to be hydrogen gas, the additional emission-bands fall far-short in contributing enough to the spectrum for white light to result from it, even to contribute enough to activate the chlorophylls.

Helium as a gas

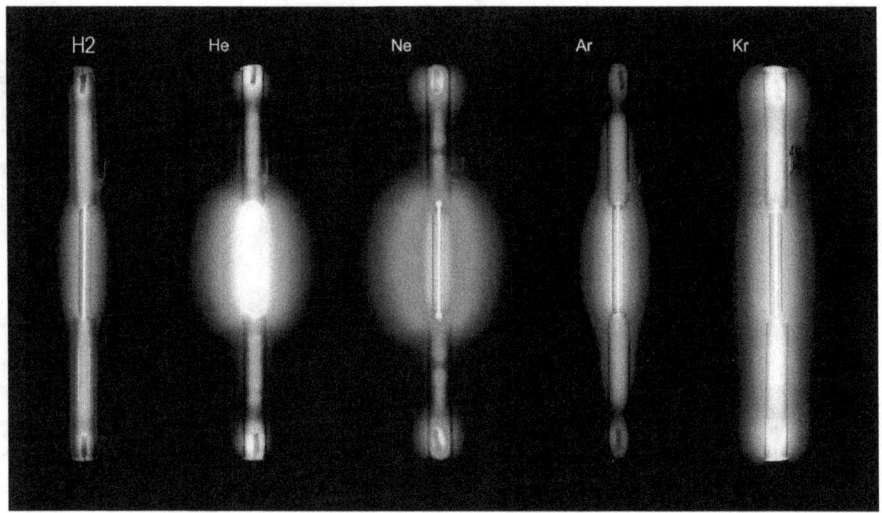

Helium as a gas, emits most of its energy in the yellow and red bands of 587 and 667 nanometres that the chlorophylls don't utilize.

The hydrogen-sun theory is totally impossible

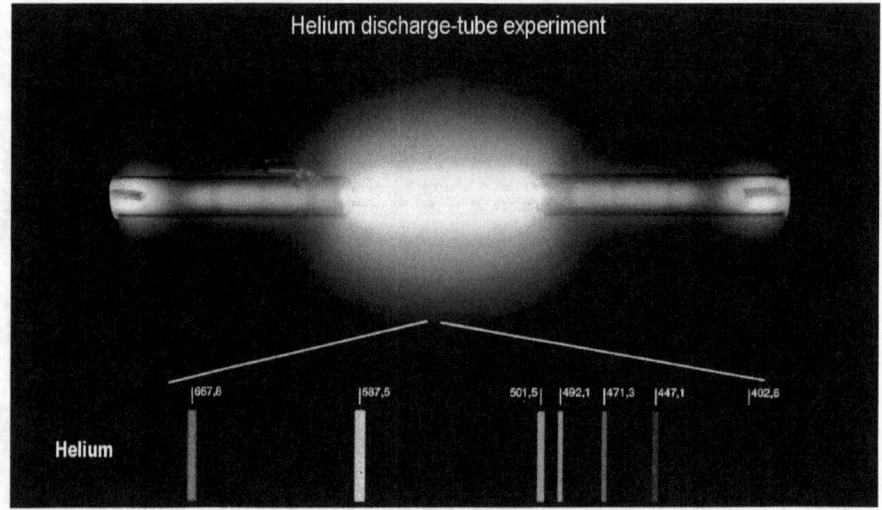

Even if all the emission bands of hydrogen and helium were combined, the result wouldn't add up to anything more than just a faintly reddish hue. This means that the hydrogen-sun theory is totally impossible as a basis for white sunlight to be produced.

In a mixture of hydrogen gas and helium gas

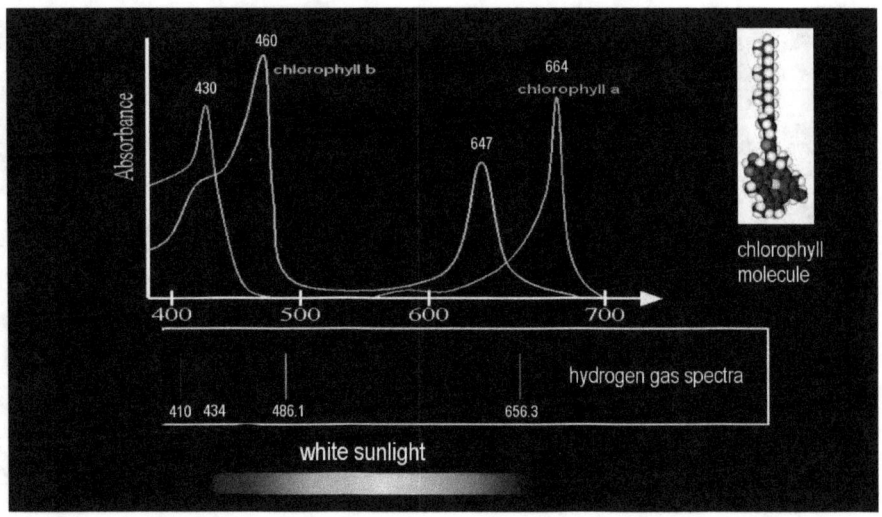

In addition, the yellow light of helium at the 587 nanometres appears in the part of the spectrum that is useless for both of the two chlorophyll types. Nor would helium play a role anyway in real terms, since none of the heavy helium atoms would be found on the surface of the Sun, under the old theory, in a gas sphere of a mixture of hydrogen gas and helium gas, as the old theory envisions the Sun, the helium gas, being twice as heavy, would congregate at the center of the Sun by the force of gravity. It would radiate from deep within, faintly, too faintly to matter.

The old theory would leave us only hydrogen light

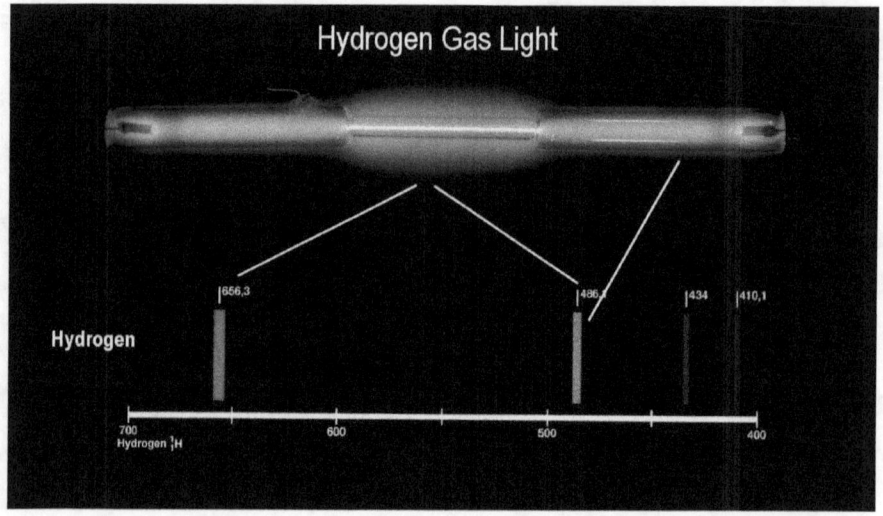

The old theory would leave us, essentially, only hydrogen light to light up the Sun, with a dim purple hue. That's far from the white light that we actually see. However, this it is all that we would get from a hydrogen sun.

Under the old theory no life would exist

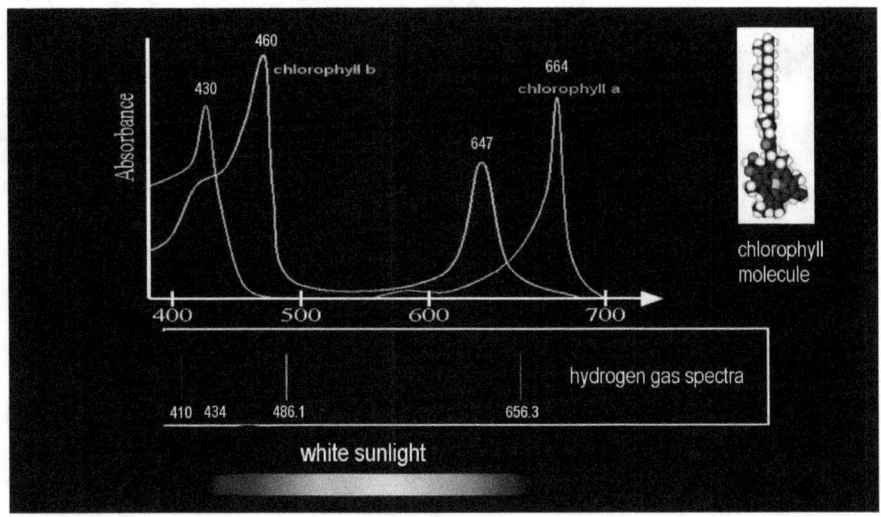

Of course, the remaining meagre spectrum that we would get from such a sun, would be far too sparse to enable the chlorophylls to function that utilize sunlight to convert atmospheric CO_2 into sugars and other nourishments that are essential for plants and plankton, and thereby for all life.

Under the old theory for the Sun, no life would exist.

Every sun would have blown itself out

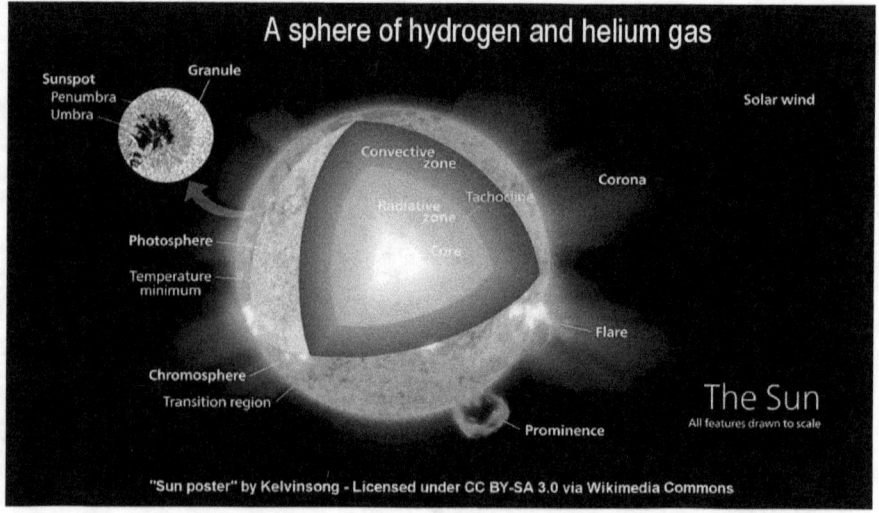

A sphere of hydrogen and helium gas

Sunspot
Penumbra
Umbra

Granule

Solar wind

Convective zone

Corona

Radiative zone

Tachocline

Photosphere

Core

Temperature minimum

Flare

Chromosphere
Transition region

The Sun
All features drawn to scale

Prominence

"Sun poster" by Kelvinsong - Licensed under CC BY-SA 3.0 via Wikimedia Commons

Actually, the Sun itself would no longer be operating. The theorized fusion of hydrogen into helium, with helium being twice as heavy, would keep the helium confined to the core of the Sun. It would clog up the fusion core with its own fusion product. The process would dilute the fusion fuel and blow out the nuclear fusion reactions.

This is what has been the universal experience with fusion-reactors in the laboratories. Every nuclear fusion reactor that has ever been built, blows itself out when the fusion product clogs up the fuel. The longest fusion burns achieved to date have lasted a mere second. No sun can exist on this basis. Every sun in the universe would have blown itself out long ago by this process, whereby the universe itself would have stopped functioning.

A new model for the Sun needs to be developed

Since this isn't the case, and the Sun is shining brightly with its color-rich white sunlight, it is fairly obvious that the assumed model for the Sun is incorrect.

Consequently, a new model for the Sun needs to be developed that stands on a platform that is demonstrably true, that is free of the many fundamental, impossible, paradoxes that render the old sun-model as something akin to the flat-earth model that fell by the wayside ages ago on the path of discovering and acknowledging what is actually real.

The path to truth is not a highway

The path to truth is not a highway at the present time. It is a difficult path for society.
Society is more-inclined to put its money into illusions that are universally accepted and politically desired, even if the illusions don't measure up, but are drab concepts that are obviously false. False concepts are often tragically destructive with dangerous consequences lurking behind the illusions, such as the global warming doctrine.

Illusions built around the epicycle

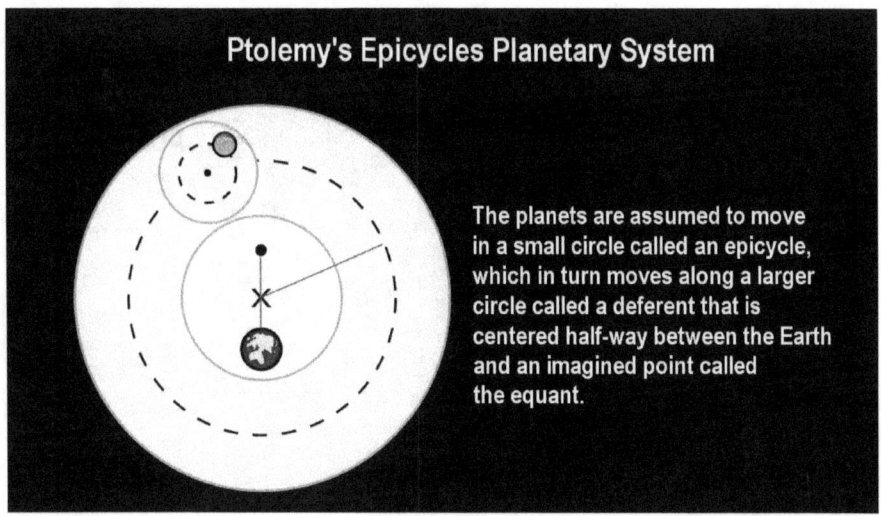

Ptolemy's Epicycles Planetary System

The planets are assumed to move in a small circle called an epicycle, which in turn moves along a larger circle called a deferent that is centered half-way between the Earth and an imagined point called the equant.

In earlier astronomy too, fundamental illusions built around the epicycle concept, had ruled for more than a thousand years, until Johannes Kepler turned the page in the 1600s with a truer perception that liberated astronomy.

The flat-earth concept

The Flat-Earth Doctrine

Engraving by Flammarion (1888)
of a traveler at the edge of a flat Earth,
who sticks his head through the firmament.

The flat-earth concept that is equally self-evidently false, had ruled even longer. It had infested the thinking of society from early antiquity on, until it faded into oblivion around the time of the Little Ice Age, though a few Flat Earth societies still linger on to the present.

The internally-heated hydrogen Sun

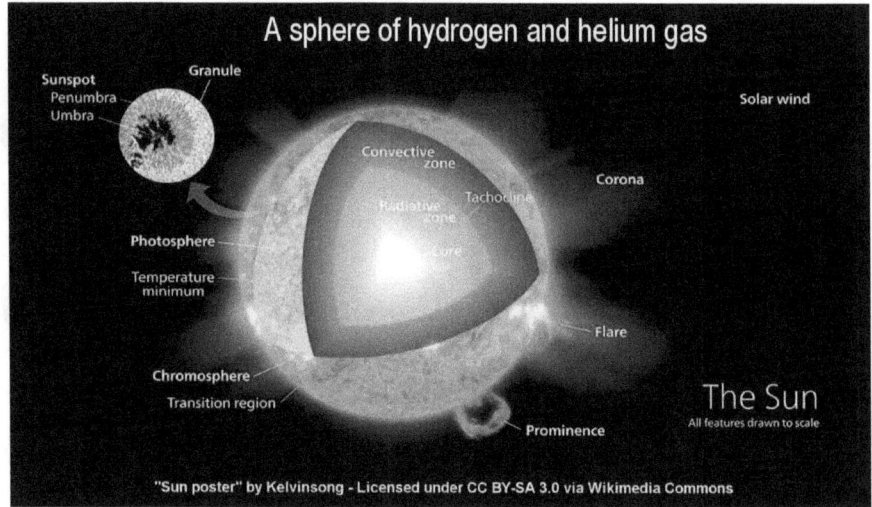

In comparison, the theory of the internally-heated hydrogen Sun that equally contradicts fundamental evidence, had only a brief history to date. Alternative concepts had actually preceded the false theory. Those were largely pushed into the background for political objectives, regardless of the physical evidence in support of them,

The evidence of the white sunlight

As the evidence of the white sunlight that enables the colorful scene that you see before you, which isn't possible under the widely accepted false model for the Sun.

The glaring evidence of a solar paradox

The glaring evidence of a solar paradox that we see all around us, suggests that the time has come to develop a new and more truthful model for the Sun and the universe, that is built on the most advanced perceptions to date, and for which a wide range of actually corroborating physical evidence does exist, which fully agrees with the evidence that we see all around us, bathed in richly white sunlight.

** A New Platform for the Sun
Building on the truth, is the most efficient principle for advancing the power of civilization.

Where is this path leading us?

It is the most-reasonable platform we can choose for creating a brighter and richer and more-secure world. So where is this path leading us, should we choose it for developing the most-leading-edge truthful perception of the Sun, and with it of the solar system, the Earth, all life, including ourselves and the Universe?

What we already know outside the box

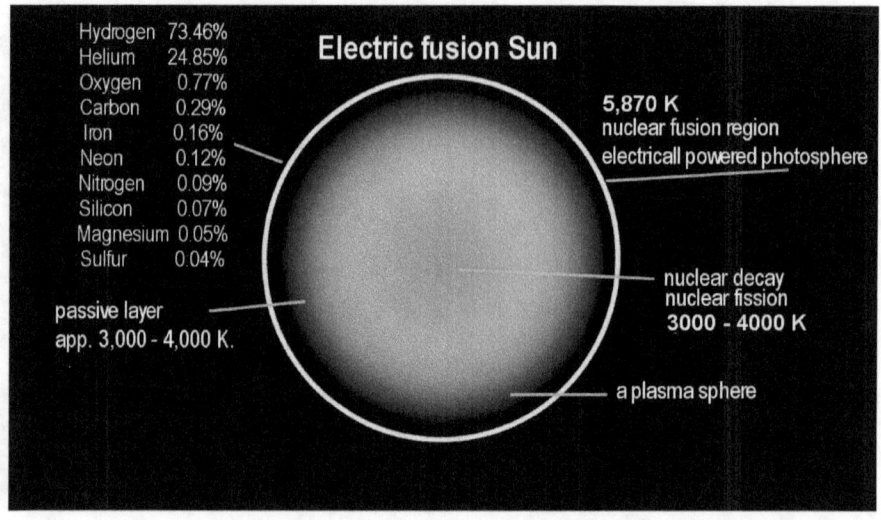

It begins with what we already know outside the box of paradoxical perceptions. We know for example that the presence of a wide range of atomic elements to be existing in the photosphere, that must exist there for white light to be generated in it, has actually already been confirmed in principle. This knowledge has been verified. It is even being utilized. We can safely build on this knowledge and move forward with it.

The atomic elements that have been recognized to be present in the solar atmosphere, are listed here. The list includes some rather heavy atomic elements, like iron. This means that those detected elements can only be present on the surface of the Sun when they are continuously produced there. Gravitational forces would have pulled the heavy elements out of the atmosphere and into the center of the Sun, as in the case of the Earth that has a core of mostly iron, a whopping 7000 Km in diameter, surrounded by a mantle of numerous lighter materials. The fact that iron is abundantly found in the solar atmosphere after over 4 billion years of the Sun's existence, with iron being 78-times heavier than hydrogen, and several times heavier than all the other elements shown here, indicates that the iron that is found at the Sun's surface and in its atmosphere, is continuously being produced by the Sun, right there at its surface, together with all other atomic elements that exist throughout the solar system. (See: http://www.lenntech.com/periodic-chart-elements/density.htm) The requirement that comes to light by the discoveries that have been made, is rather basic. The requirement is that the heavy atomic elements that are known to exist on the surface of the Sun, must be created right there, on the surface of the Sun. This requirement is easily accommodated by a Sun that is a sphere of plasma with synthesizing plasma fusion occurring at its surface. It results naturally in a process of plasma interaction that is powered externally. The principles are simple, and are supported by actual physical evidence, though the details are rather extensive.

Cold nuclear-fusion powers the Sun

I have illustrated the details of the dynamics of the externally powered Sun, that has plasma fusion occurring at the solar surface, in my video series "Outside the Box: cold nuclear-fusion powers the Sun." The scope is too wide to get into the principles involved in the exploration here, except for the one aspect that is related to the white sunlight, which is surface plasma fusion, that is actually far more straight forward than the hydrogen fusion theory is, of the old model for the internally powered Sun.

Helium nuclei named alpha clusters

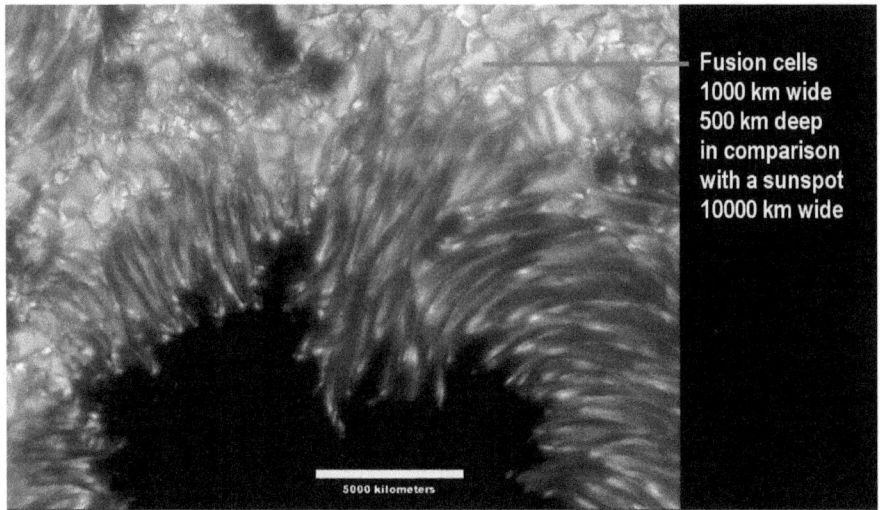

Fusion cells
1000 km wide
500 km deep
in comparison
with a sunspot
10000 km wide

5000 kilometers

In intense electric interaction of magnetically focused and concentrated electrons and protons, at the surface of the Sun, plasma reactions become so powerful on the stellar scale, that the processes readily synthesize massive amounts of the simple hydrogen atoms, and the slightly more-complex helium nuclei that appear to be so basic for everything that they were named alpha clusters.

Neutrons are created when protons join

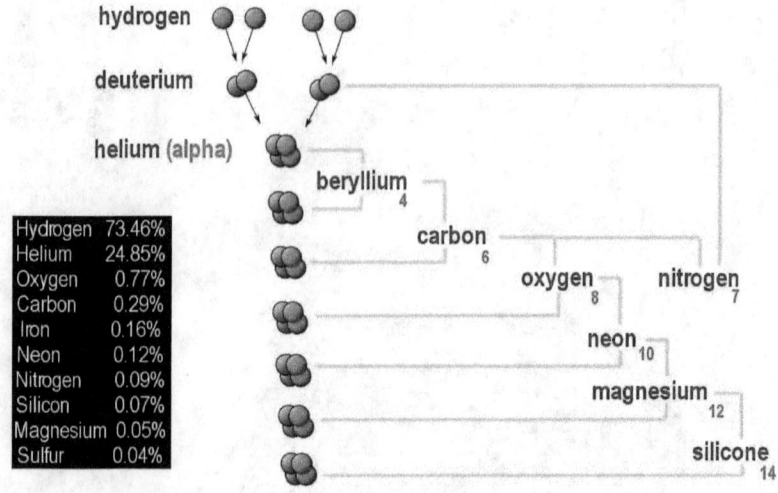

The alpha clusters consist of two protons and two neutrons. Neutrons are created when two protons join, whereby one is transformed into a neutron. It appears that in the chain of the atomic synthesizing, the alpha clusters become basic building blocks for ever-larger nuclei. The carbon nucleus, for example, could be made up of three alpha clusters, and the oxygen nucleus of four, and so on. The resulting type of chain can easily create atomic nuclei for all the elements that are known to exist in natural abundance, many of which have already been detected in the solar atmosphere. The attachment of electrons to complete the atom-building process shouldn't pose a problem either, in the highly accelerated plasma soup that the plasma-focus principles enable.

The Sun being a large sphere of plasma

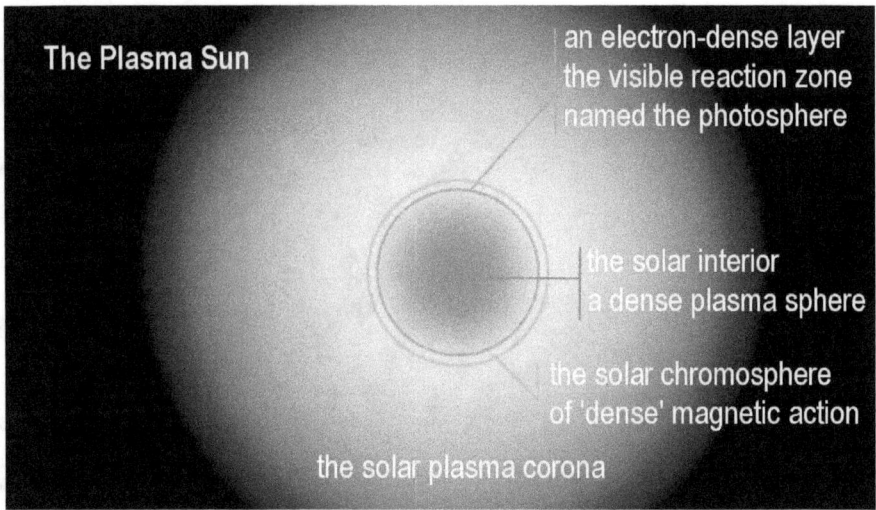

With the Sun being a large sphere of plasma, its massive gravity would force some of the plasma's electrons to migrate towards the surface. At a point, a dense layer of electrons develops that interacts almost explosively with interstellar plasma in the corona.

Termed the Primer Fields

Interstellar plasma is being focused onto the sun by large magnetic fields, which have been termed the Primer Fields by one of the explorers of cosmic magnetic phenomena. The plasma sphere surrounding the Sun, created by the primer fields, becomes the Sun's corona.

Expressed on the surface of the Sun

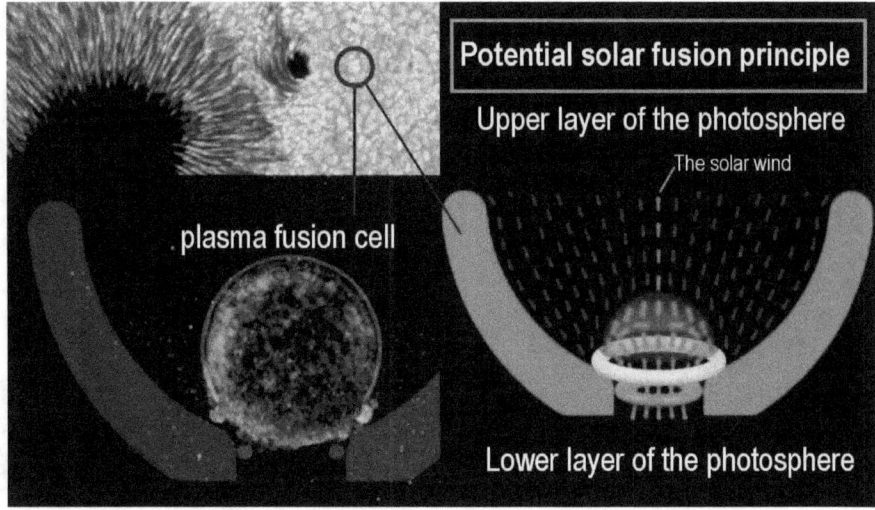

The same type of process, in principle, evidently becomes also expressed on the smaller scale on the surface of the Sun, where it 'activates' the photosphere.

The cosmic image as a model

The Red Square Nebula

By the near explosive interaction of interstellar plasma with the electron-rich layer, which takes place in the form of energized electrons and protons becoming further magnetically concentrated by the interacting processes, all add up to an immensely powerful process as the cosmic image as a model illustrates. In this immensely powerful process, that is typical for every sun, the in-flowing plasma particles are becoming fused into all the numerous types of atomic elements that are known to exist naturally. It would be surprising if this was not the case.

All atomic elements are basically structures made up of plasma particles energetically assembled.

Synthesized atomic elements

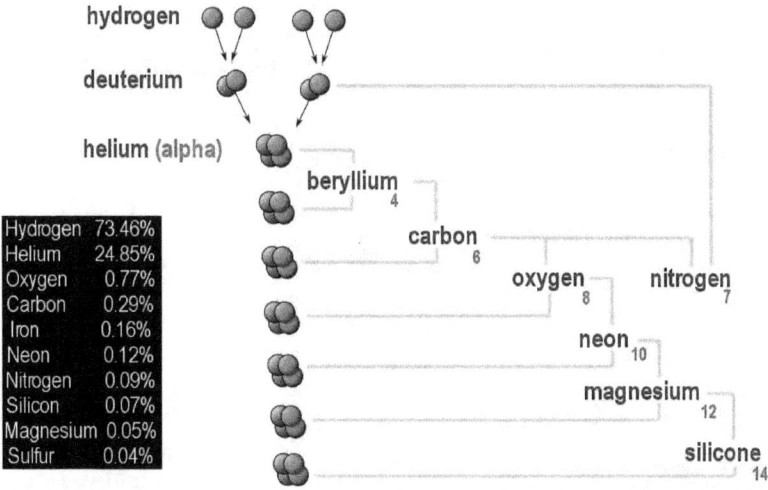

Hydrogen	73.46%
Helium	24.85%
Oxygen	0.77%
Carbon	0.29%
Iron	0.16%
Neon	0.12%
Nitrogen	0.09%
Silicon	0.07%
Magnesium	0.05%
Sulfur	0.04%

The resulting synthesized atomic elements of numerous different sizes of nucleus clusters, which result in the combining fusion-process, become electrically neutral by the attachment of a balancing numbers of electrons arranged in shells around them. By being electrically neutral, the completed atoms becoming electrically 'detached' from the plasma streams that formed them and flow away with the solar wind.

Very large atoms have a massive nucleus

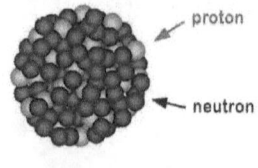

**Uranium Nucleus
92 protons, 146 neutrons**

Atomic construction principles

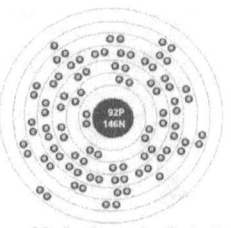

**92 electrons, in 7 shells
of orbital spaces**

Source:
http://www.saburchill.com/chemistry/
http://www.ganil-spiral2.eu/science-us/

**A shell can contain
up to 6 levels of subshells**

Very large atoms, of course, have a massive nucleus and large numbers of electrons surrounding the nucleus. Since electrons with the same electric polarity repel each other, they become arranged into separate orbital spaces that are ordered into shells around the nucleus, with the shells themselves being made up of several layers of sub shells within them. Atoms can become really complex as they get bigger, but the principles for their construction are always the same, according to the basic nature of the electrons and protons. In extremely energetic environments the plasma particles latch onto each other by the nature of the principles that govern them. The resulting atoms come in all sizes and in a wide range of different quantities. In this process all the atoms of the planets were born.
It is actually hard to imagine that such a simple type of high-energy plasma-fusing process should not be happening on every sun, if one considers the scale, the magnitude, and the energy levels that are typical for these electric processes that are operating on the stellar scale.

The magnetic compression that is achieved

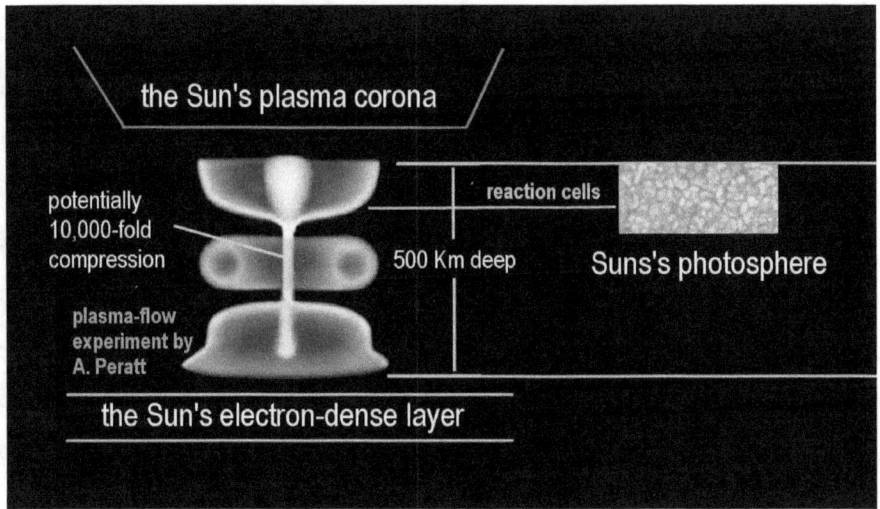

The granular nature of the photosphere, compared with the known sizes of sunspots, suggests that an average plasma fusion cell is roughly 500 km wide and 500 km deep. The potential geometry shown here has been observed in a high-power plasma-flow experiment at the Los Alamos National Laboratory. The magnetic compression that is achieved in such a process may range between 10,000 and 100,000 to 1.

A ring of up to 56 filaments of flowing plasma

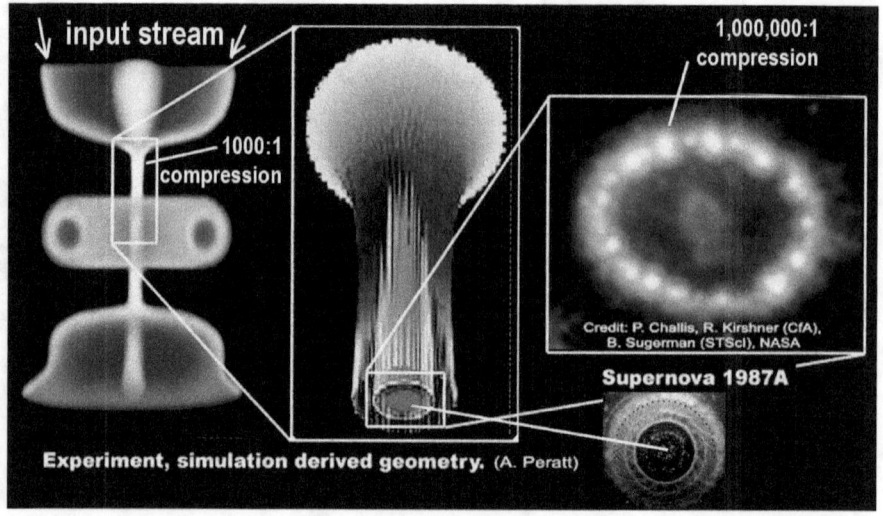

input stream

1000:1 compression

1,000,000:1 compression

Credit: P. Challis, R. Kirshner (CfA), B. Sugerman (STScI), NASA

Supernova 1987A

Experiment, simulation derived geometry. (A. Peratt)

The compressed stream has been observed in the laboratory to consist of a ring of up to 56 filaments of flowing plasma, magnetically self-aligned. Large scale evidence of this discovered phenomenon has also been observed in cosmic space.
It is hard to imagine that there should be anything that such a high-power electromagnetic particle accelerator would not be able to accomplish, operating on the enormous scale of 500 Km deep stellar fusion cells. On this scale, atomic synthesis is basically as natural as the rain that falls on Earth, even while the electron configurations for the various atoms result from extremely complex principles, especially for the larger atoms.

The most efficient principles for plasma particles to congregate

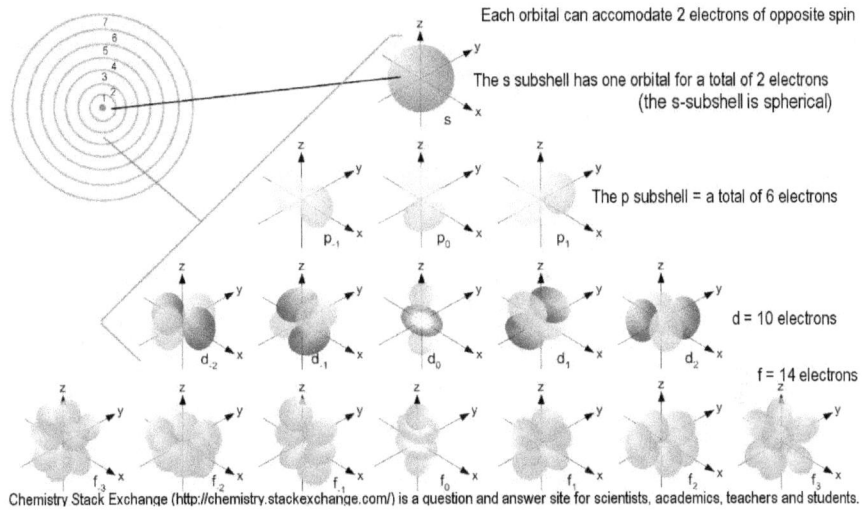

Each orbital can accomodate 2 electrons of opposite spin

The s subshell has one orbital for a total of 2 electrons (the s-subshell is spherical)

The p subshell = a total of 6 electrons

d = 10 electrons

f = 14 electrons

The construction pattern that have been discovered may be complex in their design, but they reflect in their design the most efficient principles for plasma particles to congregate closely in high-energy environments. The binding energy that is applied in the combining process, renders the resulting atoms extremely stable products.

No big-bang creation is needed

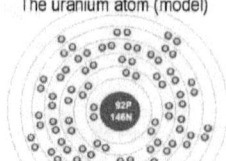

The uranium atom (model)

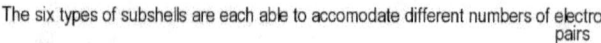

The six types of subshells are each able to accomodate different numbers of electron
pairs

The s subshell has one orbital for a total of 2 electrons
The p subshell has three orbitals for a total of 6 electrons
The d subshell has five orbitals for a total of 10 electrons
The f subshell has seven orbitals for a total of 14 electrons
The g subshell has nine orbitals for a total of 18 electrons
The h subshell has eleven orbitals for a total of 22 electrons
etc.

Each of the successive main shells can accomodate more types of subshells

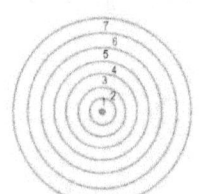

#1 shell only has the s subshell (2 electrons)
#2 shell has the s and p subshells (2 + 6 = 8 electrons)
#3 shell has the s, p, and d subshells (2 + 6 + 10 = 18 electrons)
#4 shell has the s, p, d, and f subshells (2 + 6 + 10 + 14 = 32 electrons)
#5 shell has the s, p, d, f, and g subshells (2 + 6 + 10 + 14 + 18 = 50 electrons)

The pattern is: 2,8,18,32,50... or $2n2$
In practice, no known atoms have electrons in the g or h subshells,
but the quantum mechanical model predicts their existence.

As we explore the atoms, we encounter extremely ingenuous
arrangements, but these reflect nothing more than the absolute
best way that plasma can combine into energy-forged atomic
products that by their nature can become subsequently assembled
into a near infinite variety of molecules that our world is made of.
No big-bang creation is needed for any of that. The big-bang fire
and explosion happens on the surface of our Sun continuously, and
likewise on every Sun in every galaxy throughout the universe.
The created fusion products on the stellar scale extend all the way
to uranium and slightly past that, with the smaller atoms of carbon,
iron, and oxygen being more plentiful in the product mix, and so
on.

The wide variety of atoms being synthesized

The wide variety of atoms being synthesized on the surface of the Sun, gives the emitted sunlight its full spectrum that combines into white.

The red-shift of the patterns of absorption lines

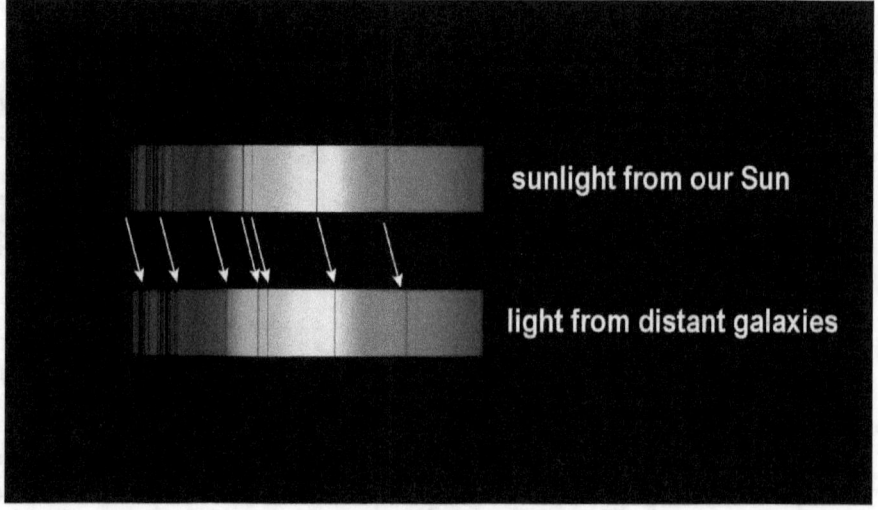

However, as I said before, when light encounters atomic elements in its path, the atoms absorb some of the sunlight, causing faint absorption lines to appear in the sunlight spectrum, according to the characteristics of the types of atoms encountered. Since the corona around our sun contains a wide variety of synthesized atoms, we see quite a few of the faint absorption lines, as we would expect to see.

It has been discovered that the same pattern of absorption lines can also be seen in light originating in distant galaxies, except that the absorption lines from distant places are shifted towards the red, as a group. What we see here, tells us that every sun in the distant galaxies operates essentially on the same principle than our own sun, and is producing the same rich white light, and is having essentially the same mix of synthesized atomic material in its corona. Obviously, the red-shift of the patterns of absorption lines is not produced by the distant object itself.

Note: The absorption lines are typically referred to as Fraunhofer lines, after the German physicist Joseph von Fraunhofer (1787–

1826) who has mapped 570 such absorption lines. The main absorption bands are caused by oxygen, magnesium, iron, sodium, and calcium, in the solar atmosphere above the photosphere. Modern observations of sunlight have detected many thousands of such lines.

Red-shift is energy depletion in the propagation of photons

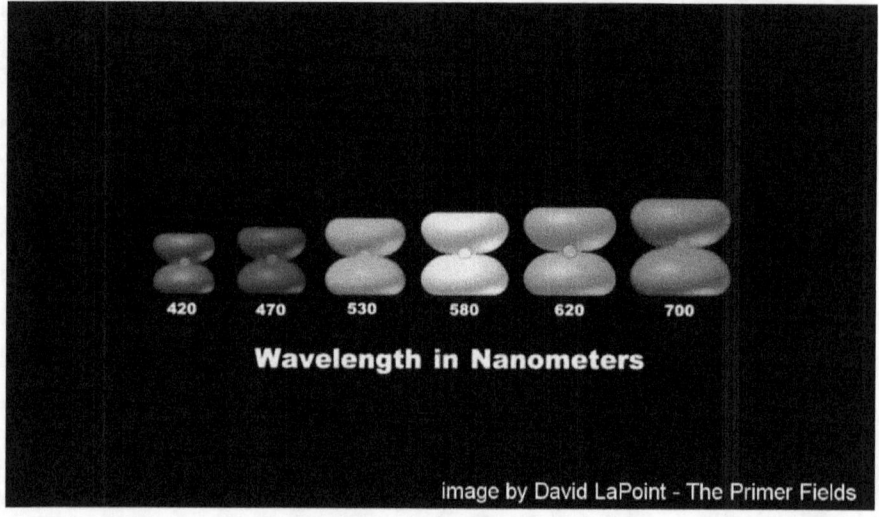

The red-shift is evidently the natural result of energy depletion in the propagation of photons over extremely long distances up to millions of light years. Energy depletion creates larger photon packets that result in longer wavelength, resulting in red-shift. The blue photon becomes a green photon, the green photon a yellow photon, and so on.

The shifting pattern as a group

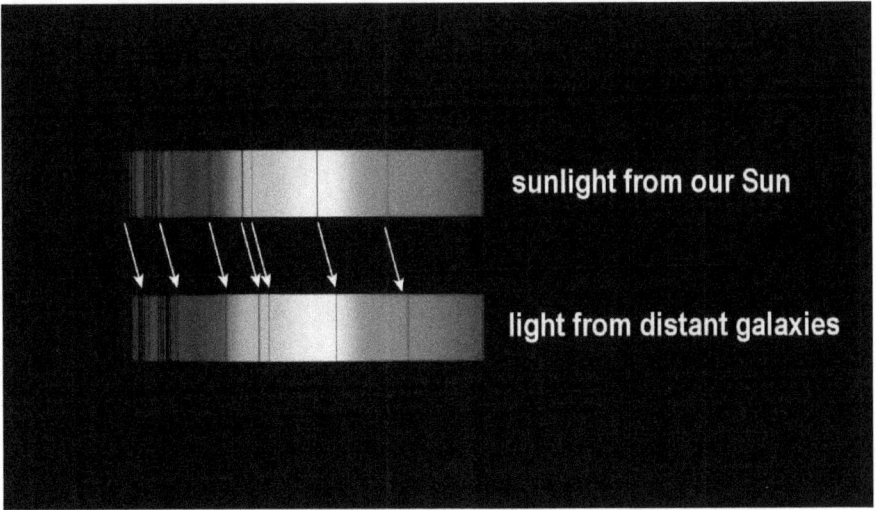

sunlight from our Sun

light from distant galaxies

The shifting pattern as a group, tells us that every sun in the universe operates in the same manner as our Sun, and synthesizes essentially the same atomic elements that are found equally present in their corona.

Obviously, the shift isn't caused by galaxies racing away from us, stretching the light, as it is still widely believed.

A universe that is not dead, and self-consuming

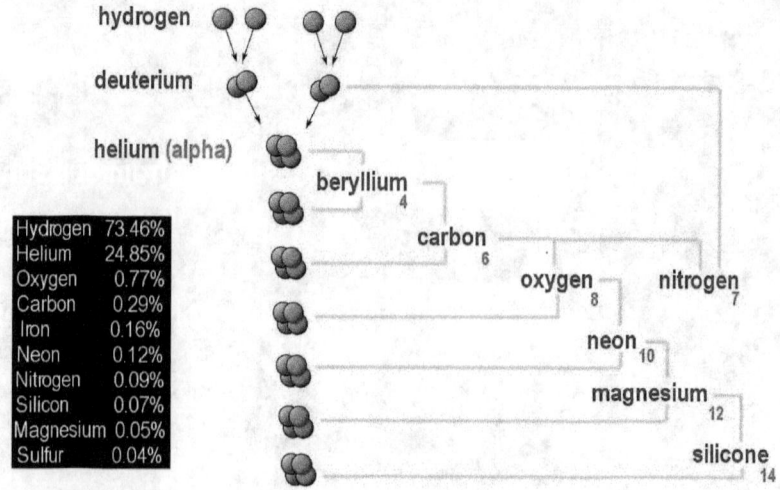

While the stellar fusion mix-ratios evidently varies, according to individual stellar intensity, the presence of large volumes of the common atomic elements in the atmosphere of every Sun, appears to be a universal phenomenon, reflecting a common, universal process. It reflects a universe that is not dead, and self-consuming as under the Old-Sun model, but is constantly self-creating, enriching, and expanding universally to the farthest reaches of cosmic infinity, and with it, it reflects the nature of ourselves as human beings and a human civilization.

Some of the atomic elements that are present abundantly in our solar corona, that confirm the plasma-fusion model of the Sun, are already being routinely utilized for exploration purposes.

NASA's Solar Dynamics Observatory satellite

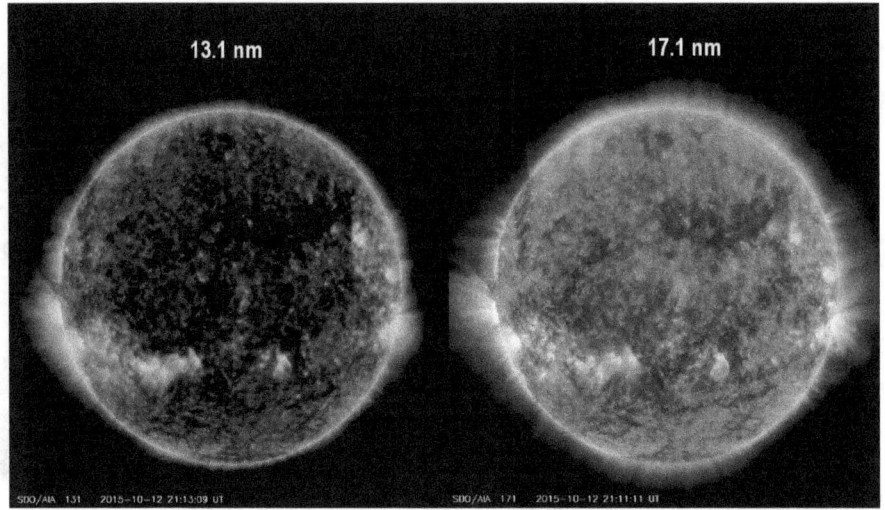

NASA's Solar Dynamics Observatory satellite, for example, utilizes the fact that some rather heavy metal-elements exist in the solar atmosphere.

Designed to look for specific bands of light

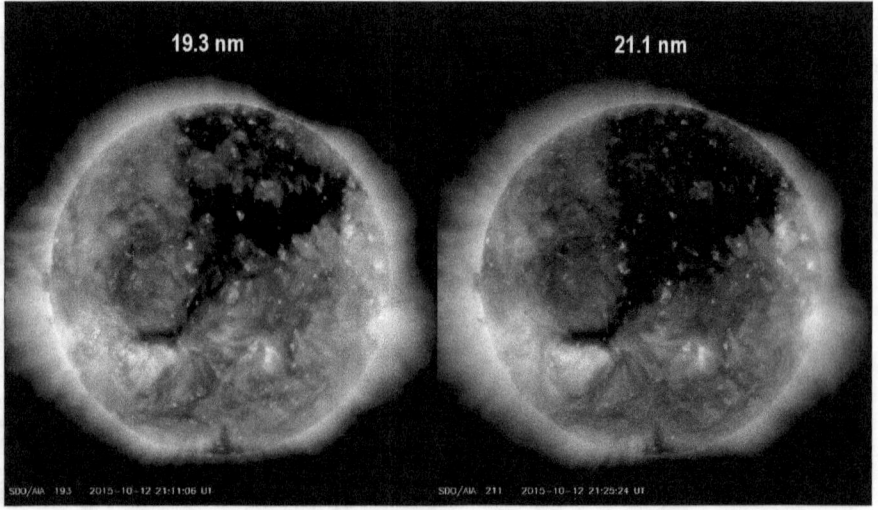

The SDO satellite is designed to look for specific bands of light that are emitted by a number of types of isotopes of iron, which emit light that is invisible to the eye, but can be seen with technology. SDO utilizes 6 different light-bands of this type for 6 of its observation channels.

The entire EUV spectrum made visible

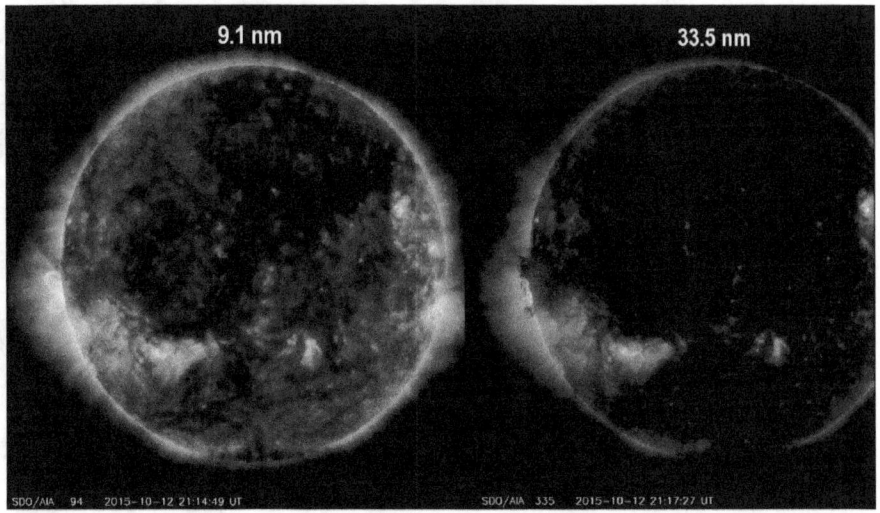

In these bands of invisible light that cover the entire EUV spectrum, the invisible has been made visible and even color-coded for us. With the 6 EUV channels that use the light from isotopes of iron, the SDO satellite explores the Sun's corona up to a million kilometers from its surface.

The presence of heavy atoms in the corona

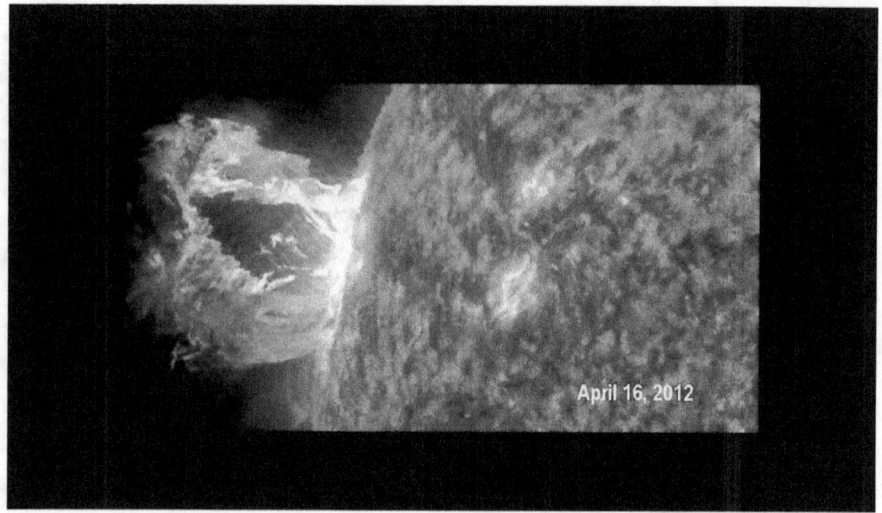

April 16, 2012

The presence of heavy atoms in the corona, even iron, evidently isn't anything extraordinary in the plasma-fusion model, so that their presence can all be reliably utilized for exploration purposes to present us images of events on the Sun that we would not see otherwise.

If the light-emitting isotopes weren't created on the Sun

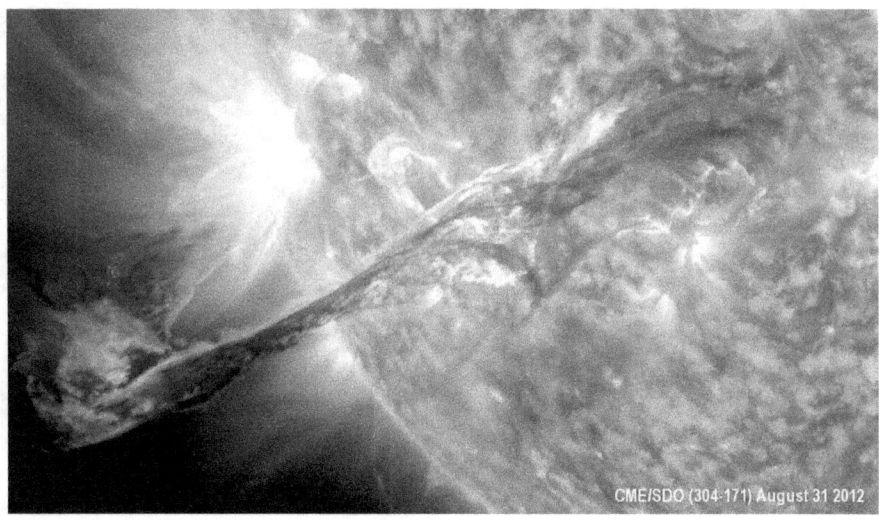

CME/SDO (304-171) August 31 2012

If the light-emitting isotopes weren't created on the Sun, we wouldn't see these amazing images that we now see.

Invisible images are made visible

NASA/TRACE - Sunspot in ultraviolet light

Some of the invisible images, that are made visible by atoms interacting with plasma near the Sun, are amazing.

The constituent elements of all the planets

It is highly likely that all the constituent elements of all the planets in the solar system have been synthesized in the high-power plasma-fusion process on the surface of our Sun, from where the numerous elements were distributed with the solar wind ever since the Sun was formed.

Synthesizing-fusion on the surface of the Sun

The highly productive plasma synthesizing-fusion on the surface of the Sun is evidently still ongoing, as it is fundamental to the emission of the white sunlight from the photosphere where all the known elements are mixed together, each emitting light in different spectra that combine into the richly white light that we receive from the Sun, which we would not have otherwise

The material makeup of the solar system

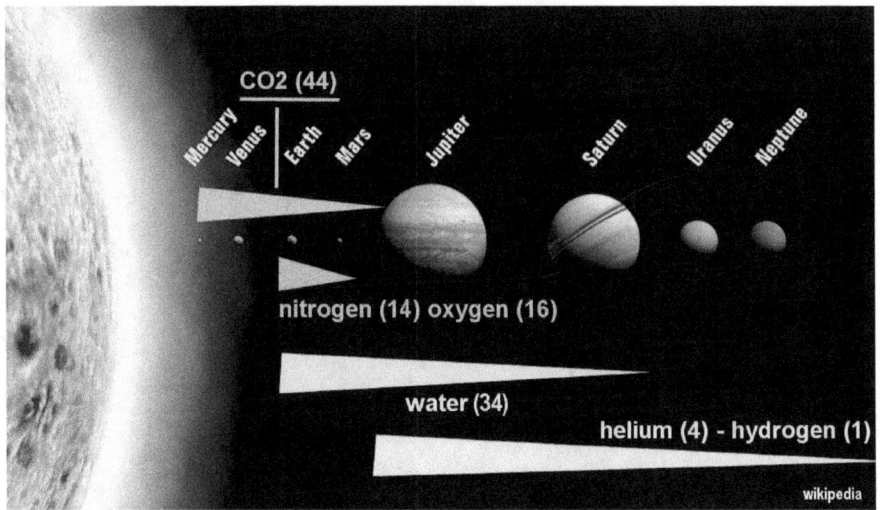

The material makeup of the solar system as a whole, does actually confirm the Plasma-Fusion-Sun concept by the fact that all the inner planets contain the heavy elements that are the first to fall out from the solar wind, while hydrogen and helium being 'lighter,' are carried further with the solar wind to form the outer planets.

Does this process appear exotic? In comparison with the Big Bang Cosmology, for which no real evidence exists, the Plasma-Fusion-Sun concept is simple, and is supported by numerous types of tangible evidence, all the way down from the makeup of the planets to the visible level on the ground in everyday living.

What stands behind the color-rich scene?

A critical part of the evidence meets our eyes every day.
Unfortunately, is what stands behind the color-rich scene, that
which makes it all possible, is a little-more challenging. For this we
have built amazing spacecraft that have opened our inner eyes.

The famous Ulysses spacecraft

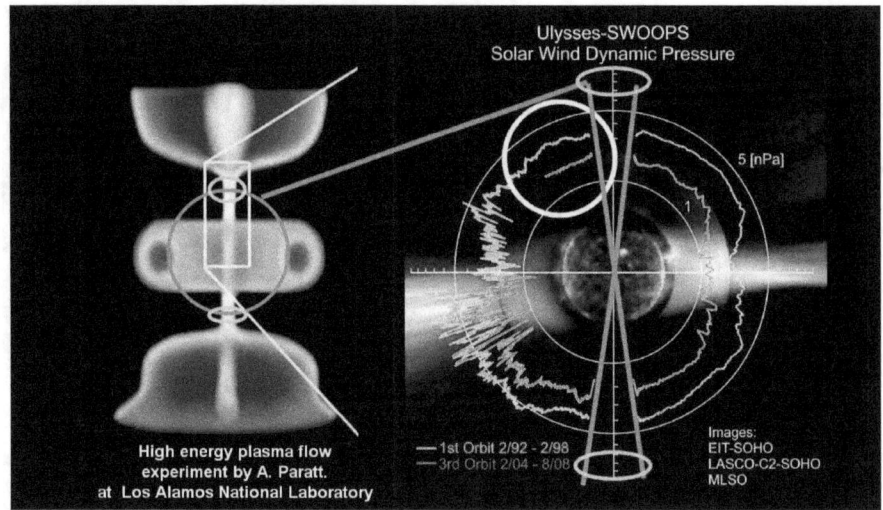

The challenge involves seeing the dynamic system that is invisible to the eye, which produces the visible evidence. The famous Ulysses spacecraft enabled us to see what powers the solar system, in the form of evidence that reflects known principles that have been discovered in laboratory experiments.

To see the dynamic system itself

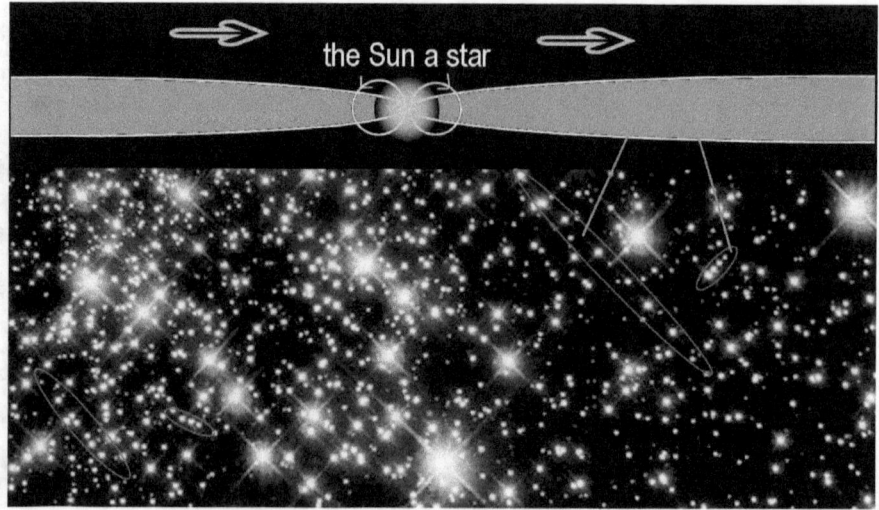

The challenge further, involves to see the dynamic system itself, as a system that is vulnerable to plasma-density fluctuations in interstellar space, which is fluctuating by its own dynamics. Here we can see the invisible by its effect in aligning stars into long electrically connected formations, like beads on a string.

Interstellar fluctuations tend to affect our Sun

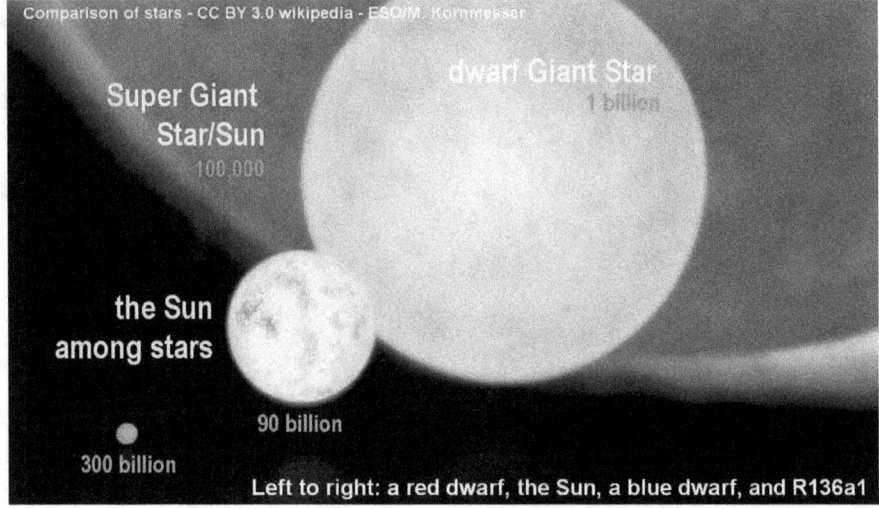

Comparison of stars - CC BY 3.0 wikipedia - ESO/M. Kornmesser

Super Giant Star/Sun
100,000

dwarf Giant Star
1 billion

the Sun among stars

90 billion

300 billion

Left to right: a red dwarf, the Sun, a blue dwarf, and R136a1

Interstellar fluctuations tend to affect our Sun more dramatically by it being a rather small and weak star, among stars.

In meeting this complex challenge of seeing the invisible, dynamics come to light that enable ice ages to occur on the Earth. Great challenges are involved in seeing the less-visible evidence behind the visible evidence.

Simple evidence that is extremely critical

This fact makes the simple evidence that is visible, extremely critical for our living at the present time, by what it represents.

What is the truth about the Sun?

Ice Age of the dimming Sun in 30 years

www.ice-age-ahead-iaa.ca

In looking behind the scene evermore deeply, we become evermore faced with the monumental question, what is the truth about the Sun and its potential for an ice age in 30 years? This question is monumental indeed, because evermore evidence points to this potential.

A new Ice Age may be erupting in 30 years

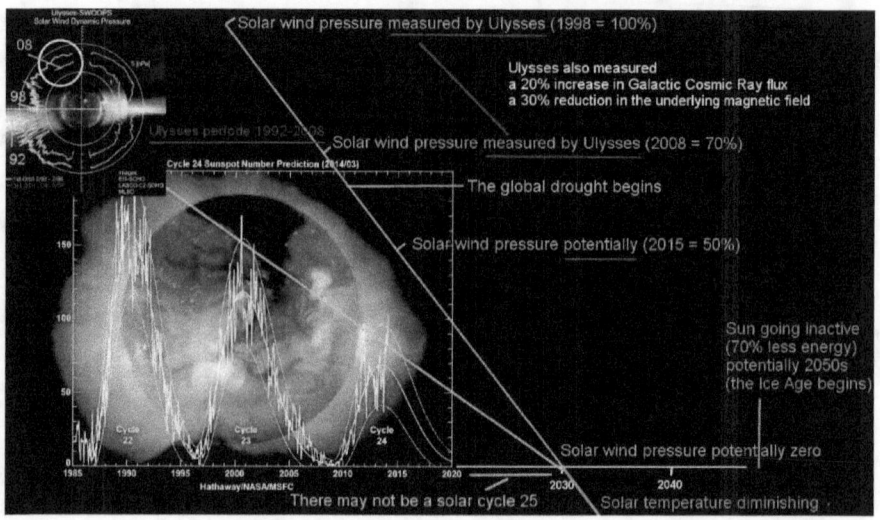

When we see evidence emerging, as we see it now, that a new Ice Age may be erupting in 30 years' time, the question, what is reliable truth, becomes fundamental to our very existence, because this question must be answered, at least sufficiently, to shape the policies of the world towards the future.

Not prepare the world for the new Ice Age ahead

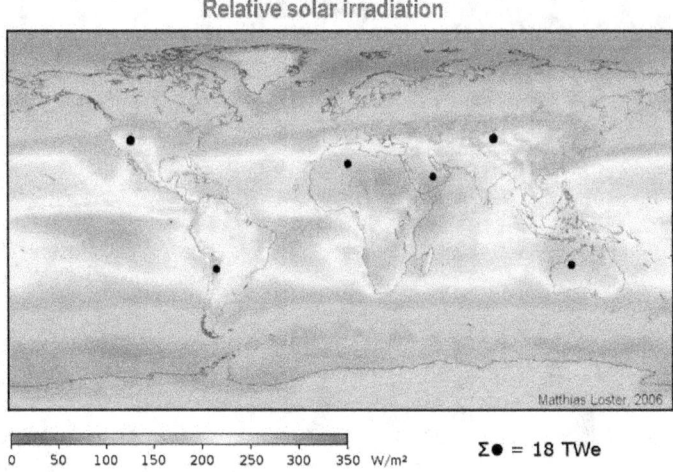

Relative solar irradiation

Should we fail on the side of doubt, and not prepare the world for the new Ice Age ahead, we would commit suicide on a gigantic scale as the result of the massive loss of agriculture and liveable territories.

Should we fail on the side of caution?

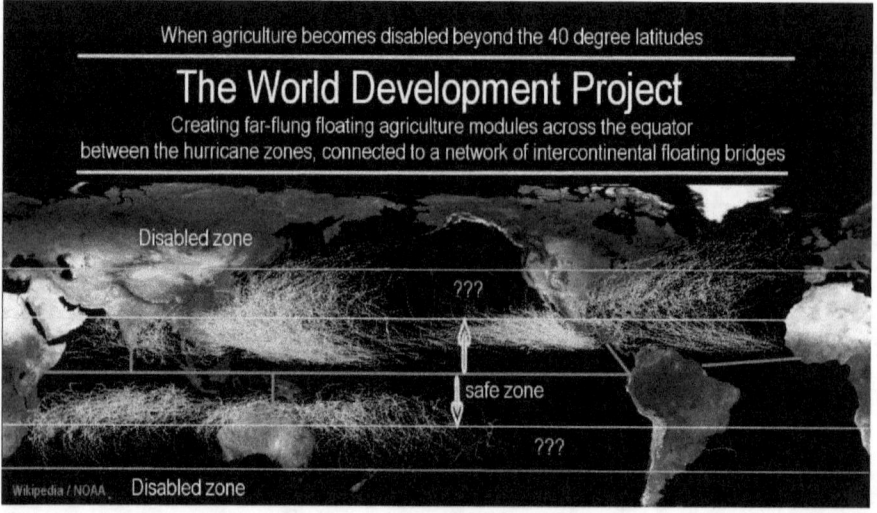

On the other hand, should we fail on the side of caution, by creating
the greatest possible industrial, scientific, and cultural renaissance
in preparation for meeting the greatest challenge of all times, we
would loose nothing if the potential ice age would never
materialize, but would gain a new and rich world for ourselves, with
the response.

The evidence that we see in the flowers

This is how we need to look at the evidence that we see in the flowers. This evidence has wide implications. It affects deeply, how we recognize the principles that affect the climate on Earth.

** Paradox of Climate Change

Our perception of the climate dynamics is presently as paradoxical as that of the sunlight based on a false theory.

Climate on Earth has been fluctuating

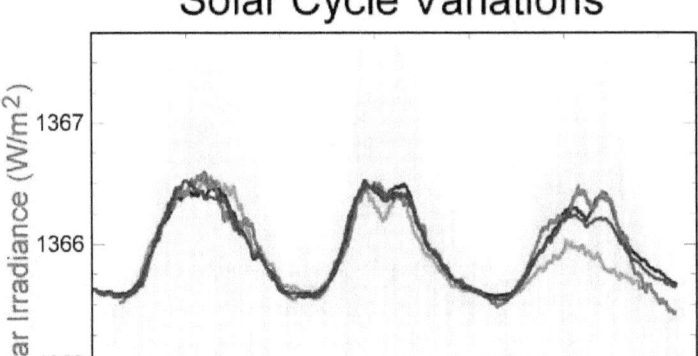

The climate on Earth has been fluctuating throughout recorded history, while to the best of our knowledge the intensity of the radiated energy from the Sun had remained fairly constant, within the range of a fraction of a percent. That's a paradox, right?
While the Sun's radiated energy does not change, we have seen wet periods and dry periods, warm periods and cold periods. We have seen dramatic changes.

The devastating Little Ice Age

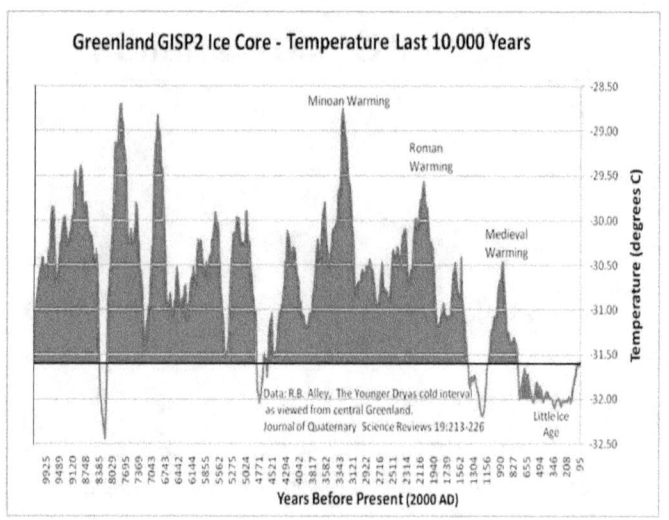

We have seen the medieval warm period collapsing into the space of a few hundred years, into the devastating Little Ice Age.

Gripped with deep cold periods

In the 1600s, the Earth became gripped with deep cold periods for a couple of centuries that decimated agriculture. Rivers and canals in Europe froze over and became skating rinks.

How was this possible? Did the Sun become colder? That's hardly possible. The measured evidence that we have on hand seems to preclude this possibility. Nevertheless, the Sun had evidently caused this massive climate change, as the Sun creates the climate on Earth in the first place. The Earth would be a frozen rock at absolute zero in temperature, if it wasn't for the Sun.

The energy flowing from the Sun to the Earth

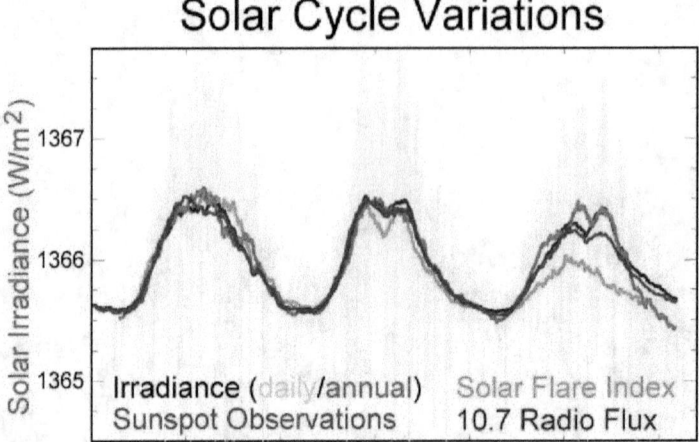

Solar Cycle Variations

The energy flowing from the Sun to the Earth amounts to a whopping 1365 Watts per square meters. The Earth is showered with this energy for 24 hours a day, constantly, with only minute fluctuations of a tenth of a percent occurring over the Sun's 11 years' cycles, termed the solar cycles?

The solar cycles are most widely known

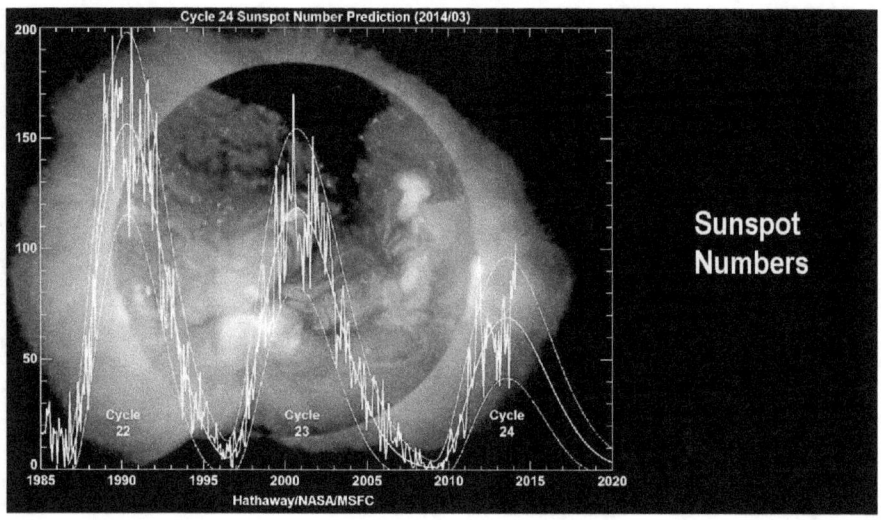

The solar cycles are most widely known for their cyclical variances in the number and the size of sunspots that occur. The sunspot number, which is a composite index, can reach to several hundred at the peak of the cycle and fall to zero during the minimum period of the cycle when at times no sunspots occur at all.

It has been recognized that these long-term cycles of the Sun, spanning 11 years on average, affect the climate on Earth, and with it affect agriculture to the point of affecting commodity prices.

Sun did change during the Little Ice Age

We have come to recognize from historic records that the Sun did change in a subtle way during the Little Ice Age and is now changing even more. Very few of the presently familiar sunspots were observed during the Little Ice Age. And in the middle of it, for several decades, no sunspots had occurred at all.

How is the climate on Earth affected by the sunspot variations?

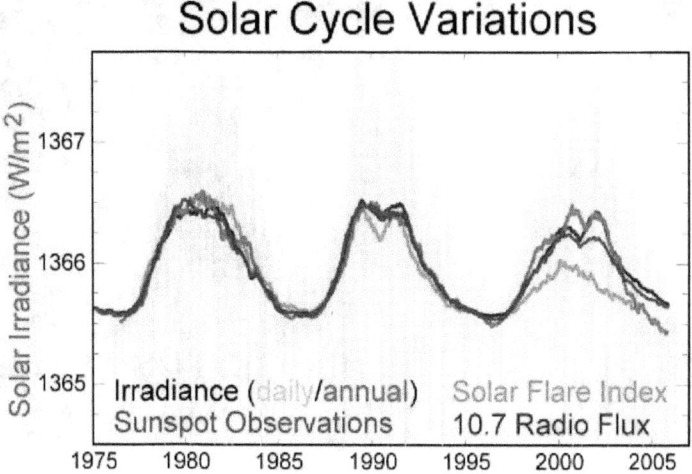

Solar Cycle Variations

But how is it possible that the climate on Earth is so dramatically affected by the sunspot variations, as in the Little Ice Age, when the radiated solar energy varies only by a tenth of a percent over the span of the solar cycles?

When we look at the Sun with telescopes

The Sun in visible light
as seen through a dark glass

In finding an answer, the technical capabilities of the space age have come to our aid. When we look at the Sun from the surface of the Earth with telescopes, through a dark glass to protect our eyes, we see only the portion of the spectrum of the lightshow around the Sun. In the visible light band, the Sun appears to us as a bright, featureless disk, except for those dark sunspots, if there are any.

The sunspots are holes ripped into the photosphere

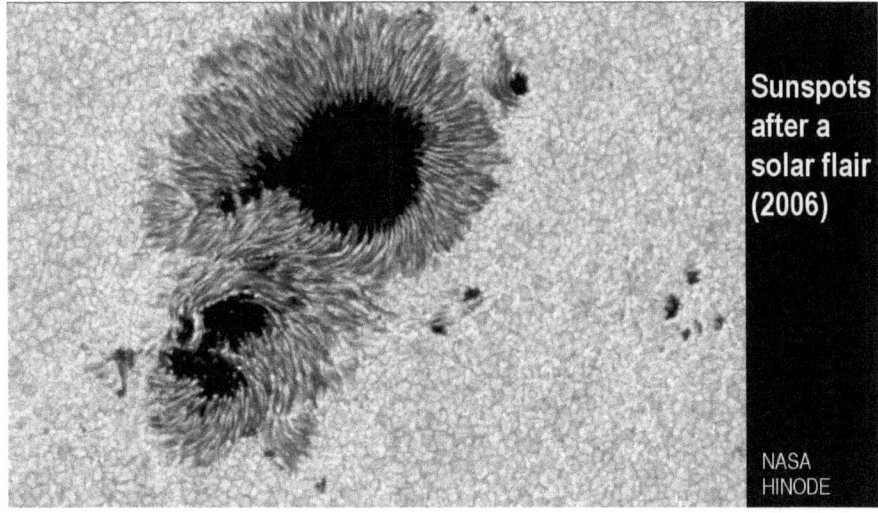

Sunspots after a solar flair (2006)

NASA
HINODE

The sunspots that we observe are essentially holes ripped into the photosphere, typically by overload explosions. Big ones can result in solar mass ejections. By looking through the umbra of the sunspots left behind, we see that the Sun is dark inside. This tells us that the energy that is radiated from the Sun, does not come from within the Sun, but from without it. Simply put, the solar energy is catalyzed by the Sun, in plasma-fusion reactions on its surface, as a by-product. The fusion process releases the radiant energy of the wide spectrum of sunlight. Overload conditions can have explosive consequences.

The Sun emits high-energy ultraviolet light

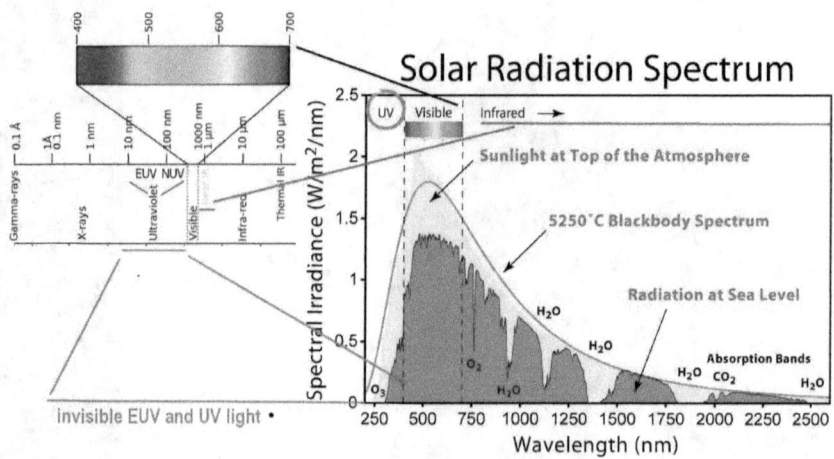

base image by "Global Warming Art images" - wikipedia

We came to recognize as our science expanded in some fields, that the spectrum of the radiated electromagnetic energy that is emitted by the Sun as light, is far wider than the narrow band of the visible light spectrum. It has been recognized that the Sun emits high-energy ultraviolet light that is of shorter wavelengths than what the eye can recognize. It has also been recognized that the Sun emits infrared light that is of longer wavelengths than what the eye can detect.

Light made up of streams of photons

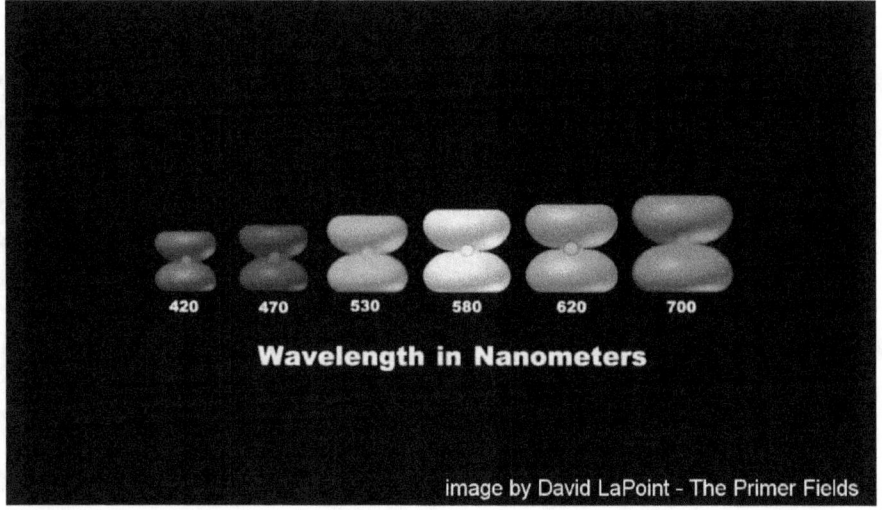

420 470 530 580 620 700

Wavelength in Nanometers

image by David LaPoint - The Primer Fields

We have come to recognize at light itself, is physically made up of streams of photons that are understood in physics as discrete packets of energy of different sizes, which we recognize as different colors. We have further recognized that the size corresponds inversely with the amount of energy contained within a packet, which holds it together. The smaller the size is, the greater is the energy it contains.

The streaming packets create waves of energy

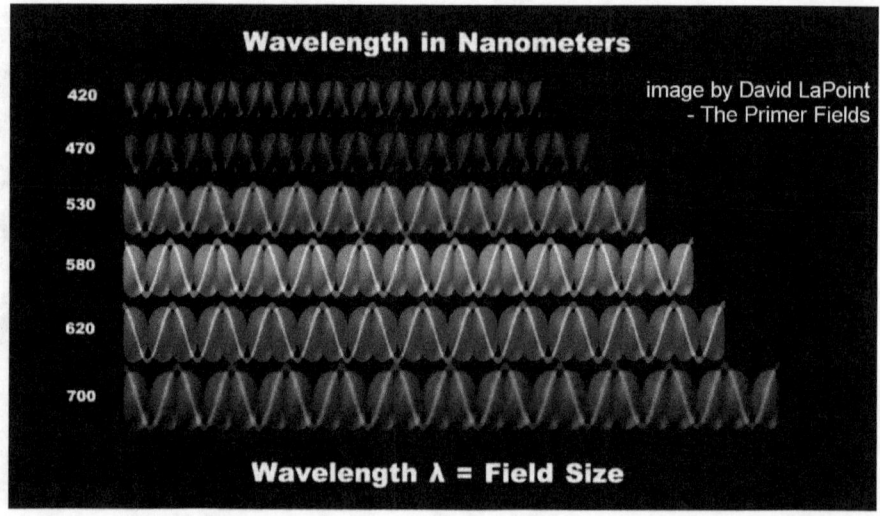

When we see light with our eyes, we see streams of packets, packets that we call photons. The streaming packets create waves of energy that have different wavelengths. The different wavelengths appear to our eyes and mind as different colors.

In the visible spectrum

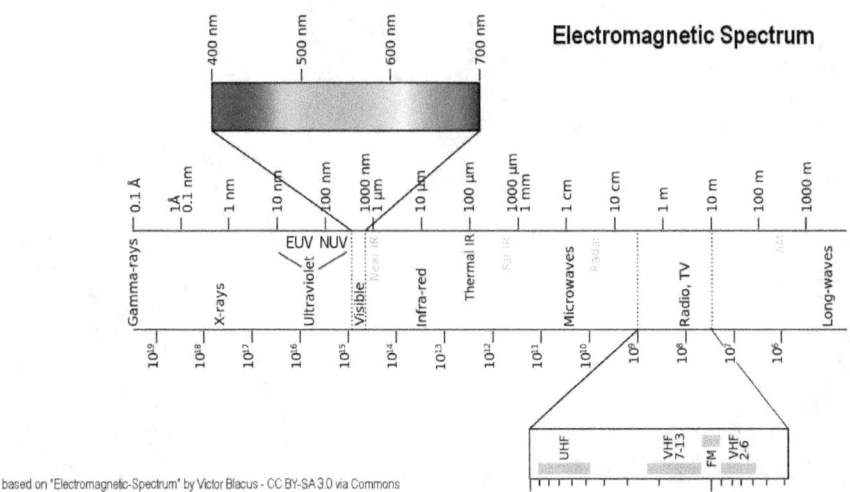

based on "Electromagnetic-Spectrum" by Victor Blacus - CC BY-SA 3.0 via Commons

In the visible spectrum, the wave lengths range from about 400 nanometres (nm) for violet light, to about 700 nm for red light. All other electromagnetic waves, those outside the visible range, can however be detected technologically. This is best done outside the Earths atmosphere, from satellites in space.

From satellites we can look at the Sun more clearly, even in x-ray light where the wavelengths are as short as just 1 nm.

When we look at the Sun in x-ray light

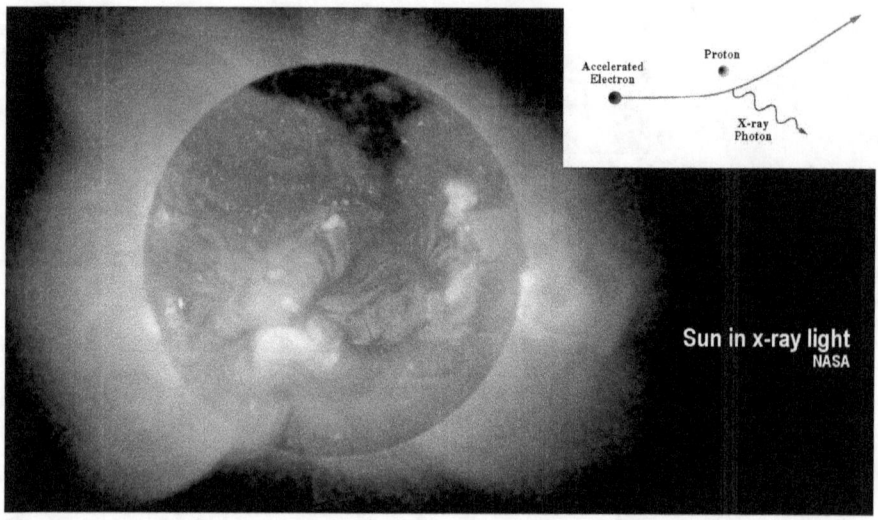

Sun in x-ray light
NASA

When we look at the Sun in x-ray light, an entirely different Sun comes into view. The Sun appears surrounded by a fog of x-ray light, and swirls of features with bright and dark areas on its surface. The intensity of the x-ray light that we see here, reflects levels of high-energy interaction of plasma particles.

When the path of a fast-moving electron is deflected by the electric attraction of it by a proton, the energy that is not deflected is split off in the form of a photon. The process is called bremsstrahlung. The energy level in plasma on the surface of the Sun and surrounding it, is so great, that a lot of x-rays are generated up to extreme distances with numerous types of features becoming visible according to local conditions. Also we see some rather astonishing results when the Sun is observed in x-ray light over long periods.

Spacecraft Yohkoh had observed the Sun in x-ray light

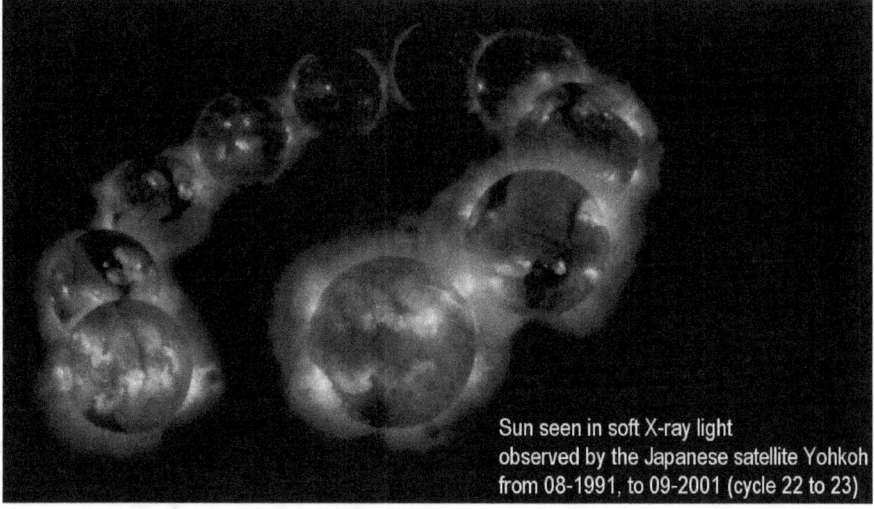

Sun seen in soft X-ray light
observed by the Japanese satellite Yohkoh
from 08-1991, to 09-2001 (cycle 22 to 23)

When the Japanese spacecraft Yohkoh had observed the Sun in x-ray light for a span of 10 years from 1991 to 2001, it was discovered that the overall intensity of the x-ray emissions varied dramatically over time in accord with the dynamics of the solar cycle. It varied immensely, by a factor of close to 100. How is this possible when the energy output of the Sun itself does not vary?

The SOHO satellite looked at the Extreme UV band

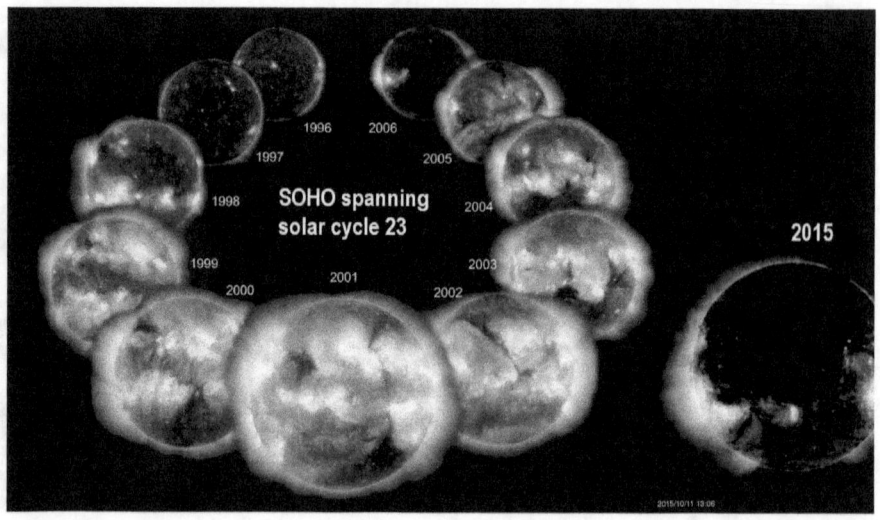

In this same timeframe a joint project was launched with international cooperation between the European Space Agency (ESA) and NASA. A satellite was deployed, termed the "Solar and Heliospheric Observatory" or SOHO for short. The SOHO satellite looked at the slightly longer wavelengths of light than x-rays. It looked entirely at the Extreme UV band. In this band too, the same patterns of dramatic fluctuations of the light intensity in the corona around the Sun was recognized, as was recorded by Yohkoh, and this too, occurred again in accord with the progression of the solar cycles. Only the intensity variation was less, probably merely 20-fold.

When one looks at SOHO's 4 light bands together

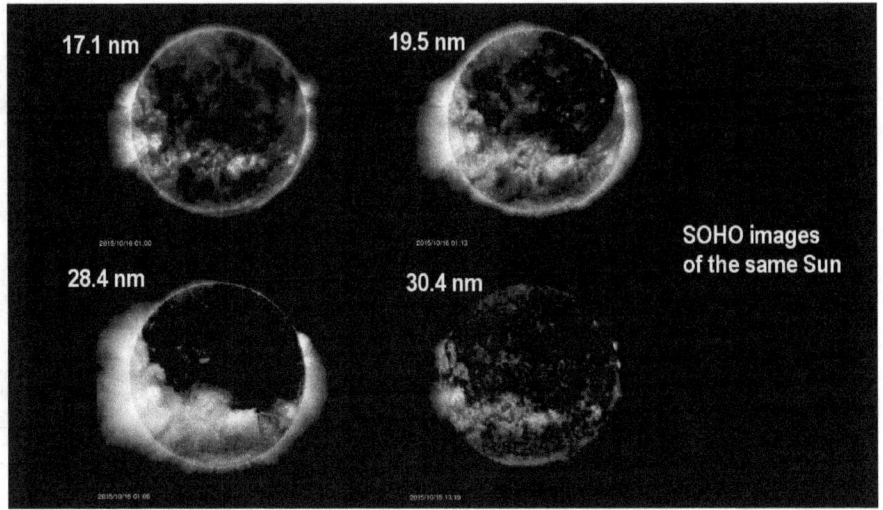

An answer to this puzzle comes to light when one looks at all of SOHO's 4 different light bands together. It becomes apparent that in the mid-range of this spectrum, in the 28 nanometer image, some parts of the Sun occasionally appear darker and sometimes appear void altogether.

When 'holes' develop in the corona

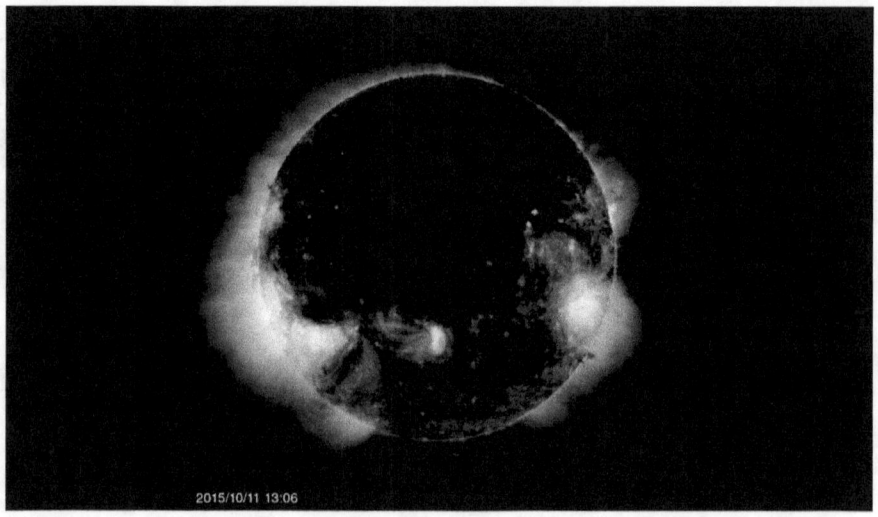

The resulting image that we see here tells us a highly significant story. We know from overall measurements that the Sun's radiated energy-level remains almost always constant within a fraction of a percent. This means that the dramatic difference that we see displayed by the satellite, reflects only peripheral differences in the surrounding solar environment, called the corona. We see the corona getting weaker so that the radiation from the Sun below, causes fewer photons to be emitted.

When the plasma is dense, the interaction of atomic elements with the solar winds cause greater amounts of photons to be emitted from the electron shells of the atomic elements that the plasma encounters in the solar wind. In contrast, when the plasma surrounding the Sun is thin, less interaction happens, or none at all. This means that the areas of lesser density can be termed coronal holes. It has been observed that the changing plasma-density around the Sun, that becomes apparent as plasma density-holes in the corona, has the potential to dramatically affect the climate on Earth.

The reason is, that when 'holes' develop in the corona, a larger volume of solar cosmic-ray flux escapes from the Sun, than would normally escape. This has big effects in the climate, even while the Sun's radiated energy remains the same.

Yes, our Sun does emit cosmic rays

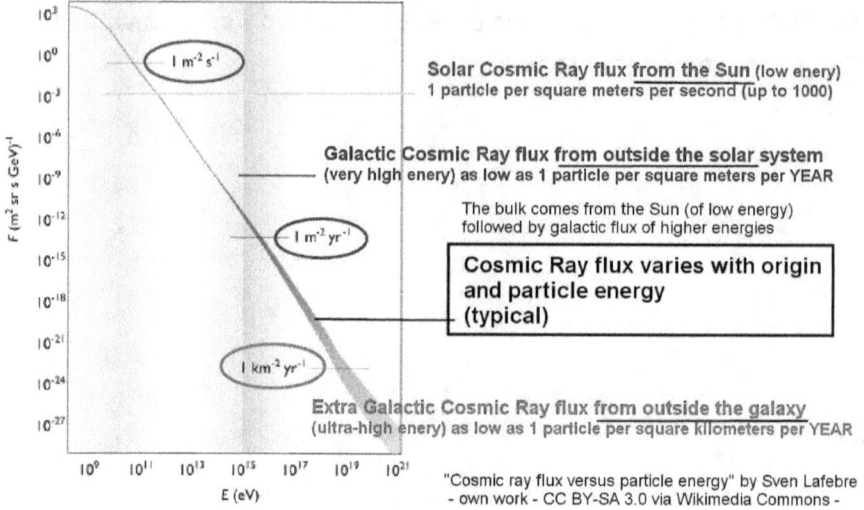

"Cosmic ray flux versus particle energy" by Sven Lafebre
- own work - CC BY-SA 3.0 via Wikimedia Commons -

Oh yes, our Sun does emit cosmic rays. Cosmic rays are not streams of light, but are single events of extremely fast moving electrons and protons with velocities sometimes near the speed of light. On our nearby Sun, large volumes of cosmic-ray emissions occur as a by-product of the plasma fusion that synthesizes atomic elements on the surface of the Sun. With the cosmic rays being emitted from the surface of the Sun, nothing much stands in their way to the Earth, except the plasma corona that surrounds the Sun. When the corona develops holes, larger cosmic-ray showers can get to us. Of course the solar cosmic-ray flux is far-less energetic than the galactic cosmic-ray flux. However, the solar cosmic-ray flux is vastly more numerous. Still, while being less energetic, the solar cosmic rays do increase the ionization of water vapor in the atmosphere. The increased ionization that enhances cloud forming, does dramatically affect our climate.

Injecting artificial cosmic-rays into a test chamber

CERN - CLOUD project - Jasper Kirkby

It has been experimentally demonstrated in principle, by injecting artificial cosmic-rays into a test chamber at the CERN lab in Europe, that cosmic-ray particles dramatically increase aerosol nucleation that is critical for the cloud forming processes.

Increase in nucleation went right off the chart

In the experiment illustrated here, the measured increase in nucleation went right off the chart when the artificial cosmic rays were turned on.

Floods of solar cosmic rays can cause floods of rain

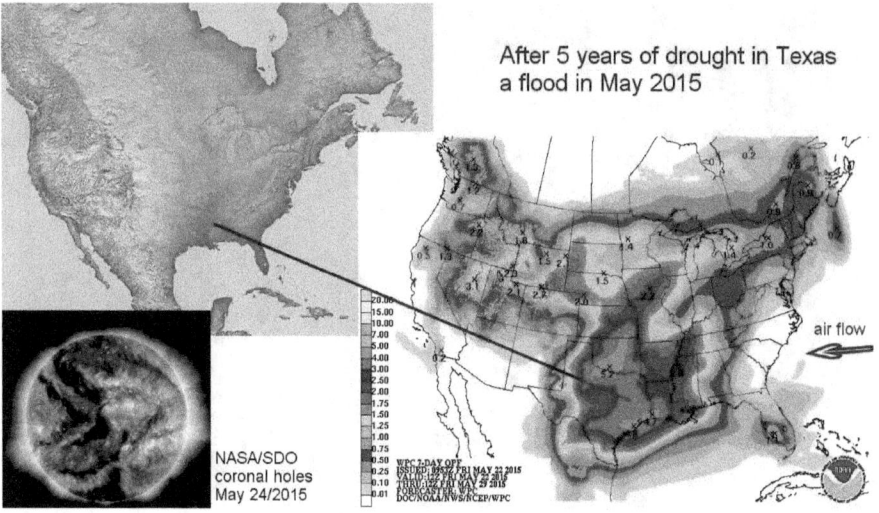

After 5 years of drought in Texas a flood in May 2015

NASA/SDO coronal holes May 24/2015

air flow

That floods of solar cosmic rays can cause floods of rain on Earth on a continental scale was demonstrated in May 2015 when wide-spread flash floods occurred from Texas to Canada in areas that haven't seen a drop of rain for years. It has been noted that this amazing event coincided with the forming of major coronal holes on the Sun.

Stronger solar cosmic-ray flux increases cloudiness

ISS-34 - Stratocumulus clouds

Typically, stronger solar cosmic-ray flux increases cloudiness over large regions, and with it affects the climate on Earth. The white tops of clouds reflect a portion of the incoming solar energy back into space. The resulting loss in solar energy for the Earth, is causing climates to become colder, even while the Sun's energy output remains the same. The cooling happens typically at the minimum portion of the solar cycles. The Little Ice Age in the 1600s evidently resulted from this type of effect when the minimum portion extended for almost three centuries.

The Little Ice Age was evidently, indirectly caused by the Sun when the Sun featured a weaker plasma corona that allowed increased solar cosmic-ray flux to reach the Earth and increase cloudiness. The cause for very-large-scale climate effects can be as simple as that.

The causative climate factor on the Earth

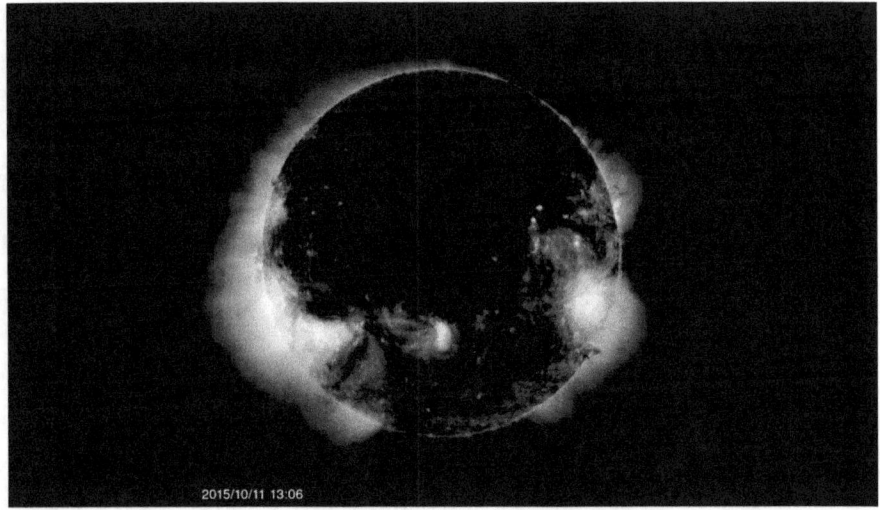

2015/10/11 13:06

The causative climate factor on the Earth, is by this clearly recognizable evidence, the fluctuating plasma density in the Sun's corona, and its gradual diminishing. The fluctuation causes solar cosmic-ray fluctuation, and with it, it affects the cloudiness on Earth that varies the portion of the incoming solar energy that is reflected back into space, which becomes lost to us thereby.

The key to the climate dynamics

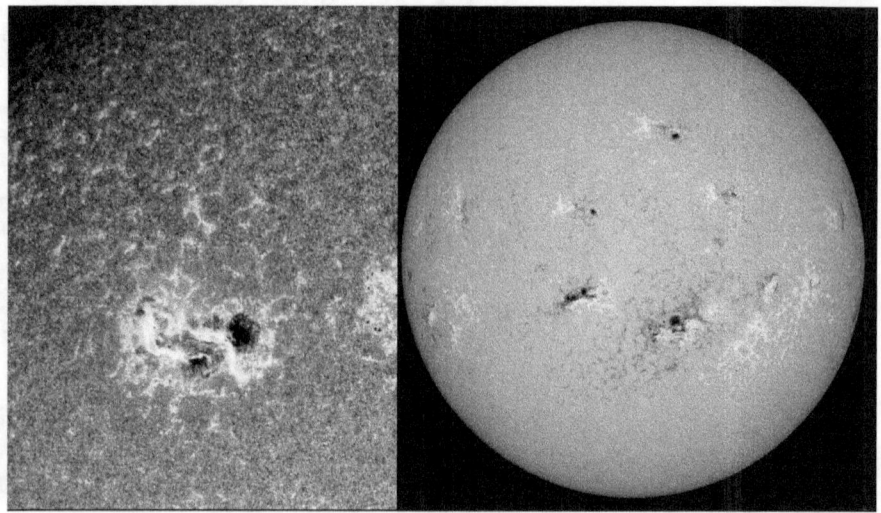

This means that the key to the climate dynamics lies on the surface of the Sun, in the plasma-fusion reactions where a large volume of solar cosmic-ray flux is generated that affects the cloud forming process on Earth.

Cosmic-Ray flux cannot originate inside of a sun

Lake Baikal, Russia
Home of the Baikal Deep Underwater Neutrino Telescope

"Jezioro Bajkal 2" by W0zny - CC BY-SA 3.0 via Commons

Solar Cosmic-Ray flux cannot originate inside of a sun. Cosmic rays are totally blocked by a few hundred meters of water. This total blocking, for example, is utilized at The Lake Baikal Deep Underwater Neutrino Telescope, that has been placed into a deep lake so that cosmic rays can't interfere with the instrument's detection of neutrinos.

Solar cosmic-ray flux provides absolute proof

The point is, that if a few hundred feet of water can block cosmic rays completely, none would penetrate the half-a-million kilometers from the core of a hydrogen Sun, that would have most of its gas densely compacted by gravity many times denser than the density of water. The existence of large volumes of solar cosmic-ray flux impacting the Earth provides another absolute proof that the widely accepted model for the Sun is fundamentally false, as it is rendered impossible by the evidence at hand. Solar cosmic rays are as impossible under the currently cherished hydrogen-sun-theory, as is the white sunlight with its full spectrum of colors that everyone sees, that under the old Sun-theory should not be possible.

The great, long-term climate changes

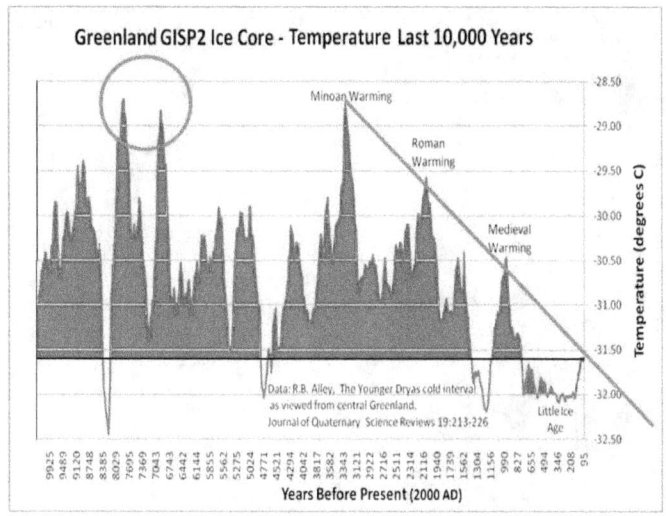

The bottom line is, that the great, long-term climate changes in recent history, that we have evidence of, such as the Medieval Warming followed by the Little Ice Age, stand as proof that the widely accepted theory of the hydrogen Sun is utterly false, as these climate changes would be impossible under this theory.

** Climate Change: A Science Paradox

That solar cosmic-ray flux is totally real

That solar cosmic-ray flux is totally real, and is a much-more powerful factor than is generally recognized, is made evident by a series of laboratory experiments conducted by a Russian scientist (Simon Shnoll). He made series of specific experiments that measured biological, chemical, and isotope decay reaction-rates repeatedly, numerous times, all in the same manner. Surprisingly, he measured different results each time.
https://larouchepac.com/20151007/new-paradigm-mankind-cosmophysical-factors-small
https://www.youtube.com/watch?v=wkDY_8HjMfk

Cosmo-physical Factors in Stochastic Processes

In a book review of his work (Cosmo-physical Factors in Stochastic Processes) on the experiments, the reviewer pointed out that when the identical experiments were conducted sequentially the results were different each time, but when several were conducted simultaneously, the results were identical in each case. Then, when the parallel experiments were repeated in the same manner, the results were different every time, but were synchronously identical. It was eventually recognized that the results of certain sequential elements spiked dramatically in 24 hour intervals.

Spiking of the experimental results in 24 hour intervals

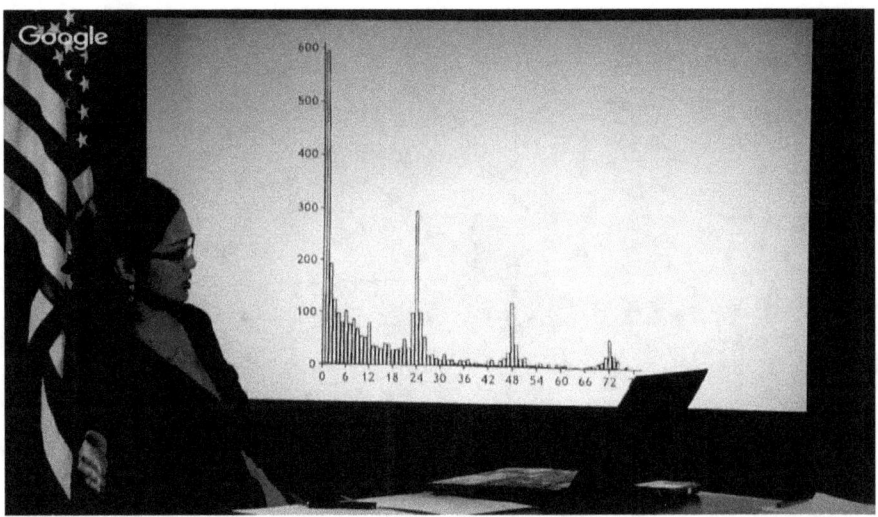

The most-likely cause for the spiking of the experimental results in 24 hour intervals would be the effect of solar cosmic-ray interaction with the experiments. Solar cosmic-ray flux spikes at any location on Earth in 24 hour intervals as the Earth rotates towards the Sun and then away from it in it 24-hour rotation. Galactic cosmic-rays do not have this precisely-shaped directionality. Galactic cosmic-rays come to Earth from random directions regardless of the time of day.

Sunspots are indicators of the state of the Sun

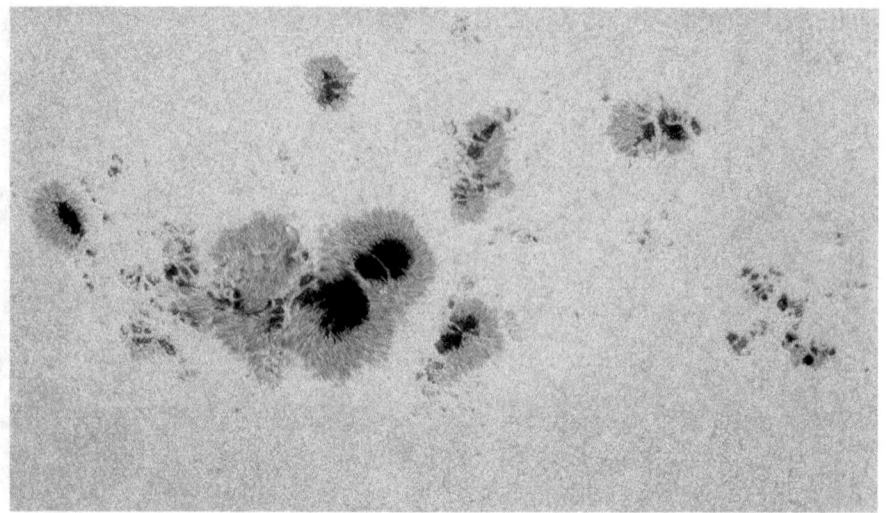

Now, what has this got to do with sunspots on the Sun?
It has been recognized that when we see lots of sunspots on the
Sun, the climate is typically warmer, and when none occur the Earth
gets colder, as during the Little Ice Age. Do the sunspots then, have
an affect on the climate, as the evidence suggests? The answer is
NO. The sunspots that we observe are merely indicators of the state
of the Sun, but are not causative climate factors. The Sun is a vast
sea of plasma-fusion reaction cells.

When the plasma-density is high

Our Sun is a sea of vast electric current streams in motion

NOAA 11364 Dec 01, 2011 11:00 UT Zeiss AS200/3000
Lumenera LU075M 3000 frames Lunt LS50 BF1200 h-alpha filter

Rogerio Marcon
Campinas SP Brasil

When the plasma-density is high, solar activity is high. Overload conditions (sunspots) can occur in this strong environment.

Sunspots result from regional overload conditions on the Sun

Solar activity and sunspots

NASA - TRACE image 17.1 nm

Simplistically speaking, sunspots result from regional overload conditions on the Sun. When the fusion products are not vented fast enough, or too much back-pressure develops, some reaction cells become 'clogged up' and disintegrate.

When they (the fusion cells) cease to function

Electric Solar Activity

When they (the fusion cells) cease to function, the plasma pressure stops and the cells blow out.

When solar mass-ejections occur

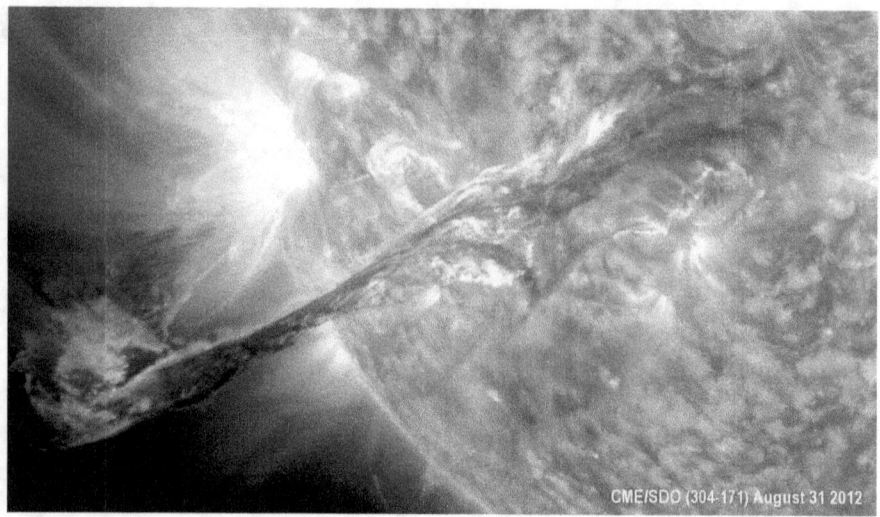

CME/SDO (304-171) August 31 2012

When they blow out in a big way, solar mass-ejections occur. The overall effect on the Earth's climate, however, is too minuscule to be notable.

No matter how big the mass ejections may become

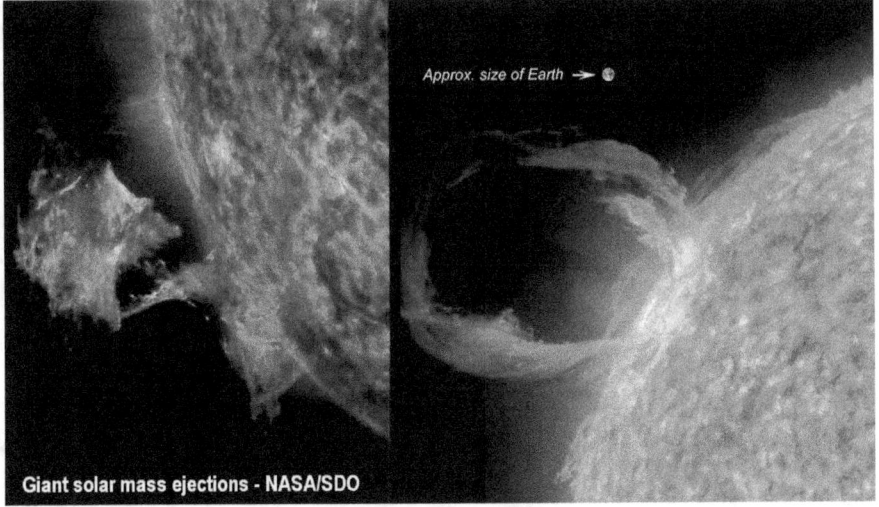

Approx. size of Earth ➜ 🌑

Giant solar mass ejections - NASA/SDO

No matter how big the mass ejections may become, even dramatically big, as they become in some cases, their duration is too short, their numbers are too low, and their energy is actually too puny to alter the overall climate on Earth. But they do have the potential to cause earthquakes.

What affects the climate on Earth

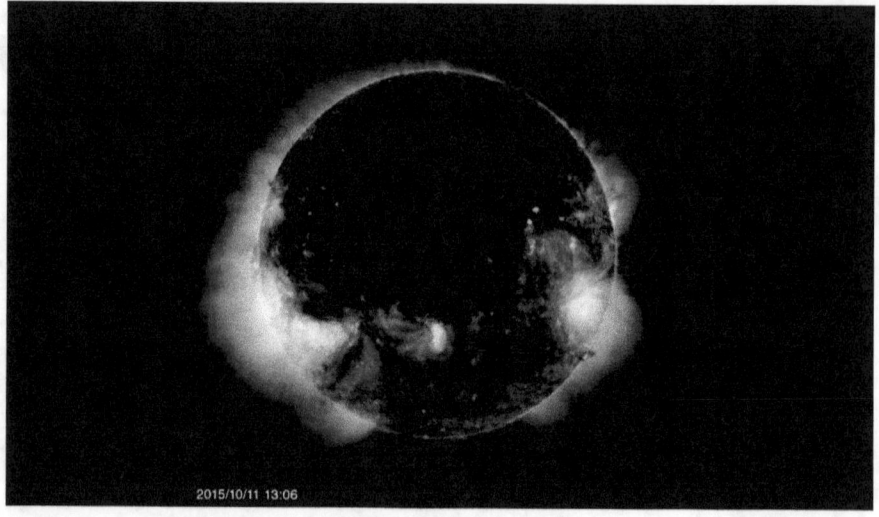

2015/10/11 13:06

What affects the climate on Earth is the changing density of the plasma surrounding the Sun. When the corona density is low, which coincides with fewer sunspots, fewer solar cosmic-rays become trapped in the corona. The resulting increase in solar cosmic-ray flux has a major affect on our climate, by affecting global cloudiness. The cause for long-term climate changes becomes evident by the long-term trend in the number and the size of the coronal holes that occur. The changing cloudiness and water-vapor density in the atmosphere reflects this trend.

High-intensity Solar seasons

View of the Earth from ISS, Jan.4 2013, from over the mid-Pacific, 460 miles east of northern Honshu, Japan.

Changes in the density of the solar corona that traps the solar cosmic rays to varying degrees, cause changes in cloudiness. The high-intensity Solar seasons that we experience as warmer climates, do not result from the Sun giving us more heat, or the greenhouse gases absorbing more heat. Instead, the warmer climates are caused by fewer clouds that reflect less sunlight back into space, and clearer skies that allow more sunlight to reach the surface. In addition, less cloudiness results in a stronger greenhouse effect.

When fewer clouds form

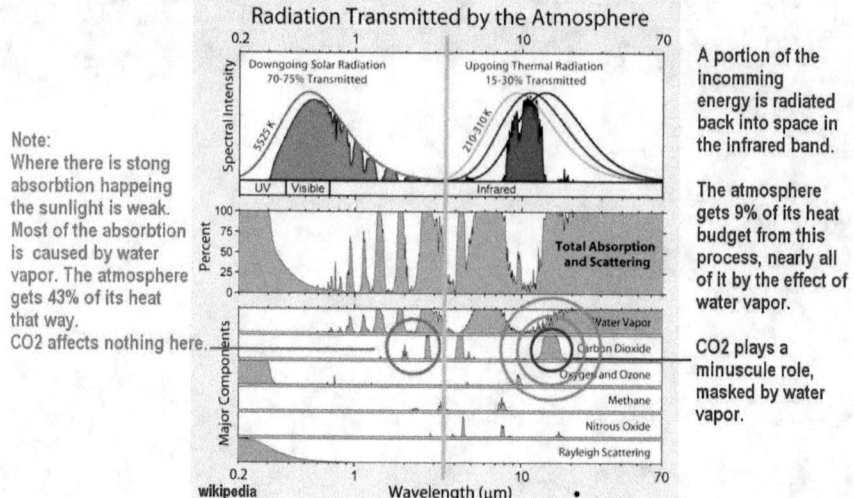

Radiation Transmitted by the Atmosphere

Note:
Where there is stong absorbtion happeing the sunlight is weak. Most of the absorbtion is caused by water vapor. The atmosphere gets 43% of its heat that way.
CO2 affects nothing here.

A portion of the incomming energy is radiated back into space in the infrared band.

The atmosphere gets 9% of its heat budget from this process, nearly all of it by the effect of water vapor.

CO2 plays a minuscule role, masked by water vapor.

When fewer clouds form, more water vapor remains in the atmosphere, which has a major effect. Increased water-vapor density in the atmosphere increases the greenhouse effect, because, up to 97% of the atmosphere's greenhouse effect is generated by water vapor. Oxygen contributes a bit. And some of the greenhouse effect is caused by the Rayleigh Scattering effect. Carbon gases have no comparable effect at.

The carbon greenhouse gases

The carbon greenhouse gases, such as CO2, affect nothing on the climate front. They contribute far too little to be significant. Their contribution to the greenhouse effect adds up to a millionth part of it.

The giant factor in the greenhouse equation, is first and foremost, the density of water vapor. This density is critically affected by solar cosmic-ray flux, and to some degree also by galactic cosmic-ray flux, which too, is increasing. Paradoxically, this giant factor that outweighs everything and is supported with real evidence, is completely ignored in what is termed "climate science." The ignorance is so deeply cutting that the only factor that is allowed to be considered for climate change in the world, is human activity, which ironically has no effect at all. 10% of a millionth part in the greenhouse equation, which the manmade CO2 does contribute, is too little to be significant in the face of the giant factors. I have presented some of the details of the CO2 myth in a video presentation with the title shown here.

In climate science the cosmic-ray factor is ignored

We face the same paradox therefore, in what is called climate science, where the cosmic-ray factor is staunchly ignored in support of a false model, just as we face the hydrogen-spectrum paradox in cosmology that is ignored in support of the same false model. The results are tragic in both cases.

It would be wonderful if humanity had the power

Corel corp.

It would be wonderful if humanity had the power to affect the climate on Earth. It would be able to avoid the ice ages, which too, are caused by what affects our Sun.

With the next Ice Age being only 30 years distant

That's why the climate story can't be allowed to end at this point, because the real causative factors that affect the Sun are also the factors that cause the ice ages, with the next Ice Age being only 30 years distant, potentially.

** Absolute Climate Change: A Science paradox

When we speak of an Ice Age

The Reverend Robert Walker
Skating on Duddingston Loch,
attributed to Henry Raeburn, 1790s

When we speak of an Ice Age - and I mean the actual, real Ice Age - the climate transition promises to be nothing small on the scale that we have experienced before.

It won't even be an environmental factor

It won't even be an environmental factor that causes this change.

We are facing an Absolute Climate Change

Ice Age of the dimming Sun in 30 years

www.ice-age-ahead-iaa.ca

We are facing an Absolute Climate Change that has not been experienced before in recorded history. We are looking for absolute factors to cause this absolute change.

The Little Ice Age in the 1600s

The Little Ice Age in the 1600s had caused havoc and spread starvation across Europe when agriculture failed. It had killed 10-30% of the population with famine.

The Little Ice Age was a minuscule affair

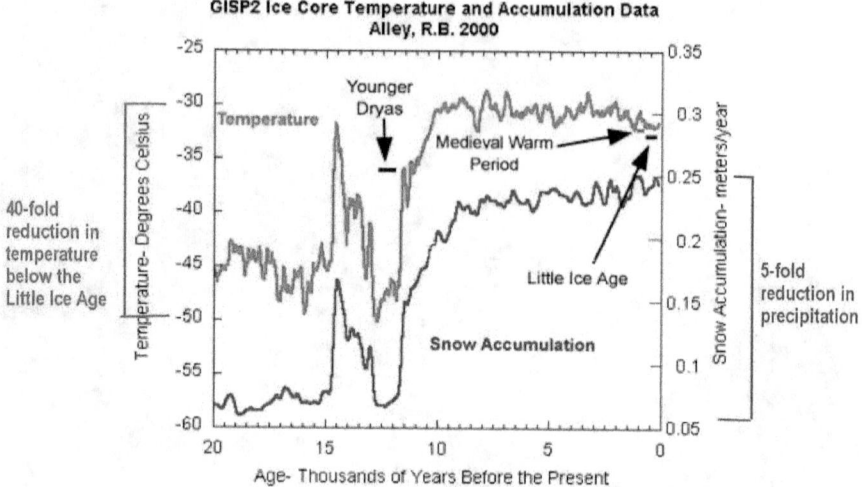

Nevertheless, the Little Ice Age was a minuscule affair that is barely recognizable on the big scene. It gave us a half of a degree in global cooling. This stands in comparison with the 20 degree cooling that the next Ice Age promises, according to historic ice core data. This is what Absolute Climate Change looks like.

This absolute climate change won't be caused by cosmic-ray-factored changes in cloudiness. The driving factor in this case, will be the Sun itself, with the Sun going inactive altogether, or nearly so. Nothing less can cause a 40-times colder world than the Little Ice Age had been.

*The Sun as an inactive star

The Sun, of course, will continue to exist, as an inactive star, and radiate energy in some form, from its residual stores, or from low-level nuclear fission energy being produced within it.

For this absolute change

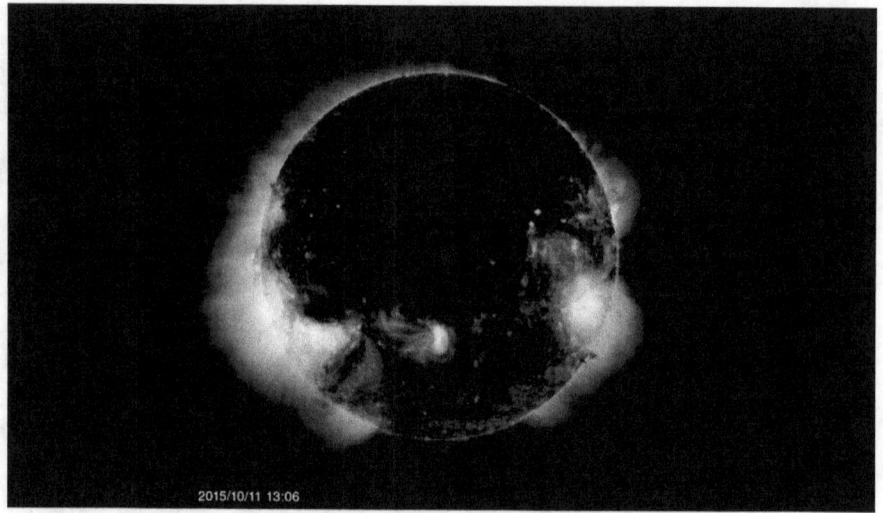

2015/10/11 13:06

For this absolute change, the process that currently causes the corona plasma fluctuations and with it coronal holes, will undergo an absolute change.

With the Sun being externally powered

The Red Square Nebula

With the Sun being externally powered by interstellar plasma streams that are electromagnetically focused onto the Sun, as in the example shown here as a model, we are looking at fragile dynamics that operate with built-in positive feedback at every stage.

A positive feedback system is one that operates wonderfully stable for as long as all the critical conditions are met. However, such a system is prone to chaotic collapse when the required conditions diminish below a critical minimal threshold.

Rapid transitions and chaotic collapse

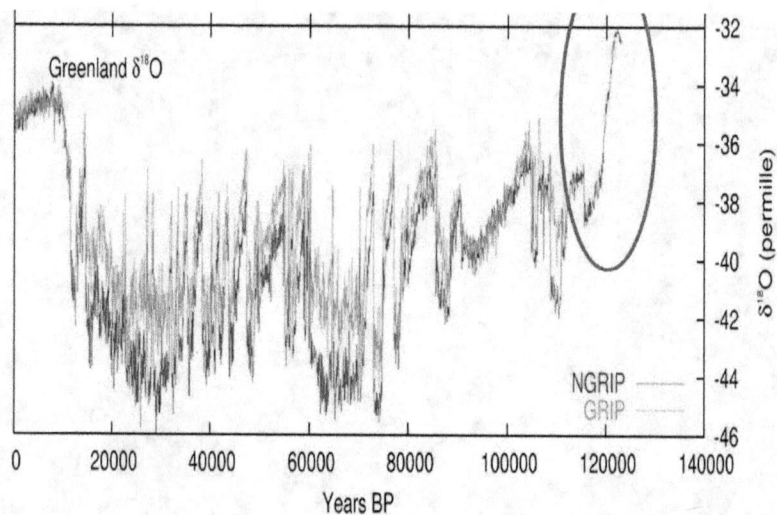

Rapid transitions and chaotic collapse - as we see it illustrated here for the start of the last Ice Age - are typical for all positive feedback dynamics when critical thresholds are crossed. The solar system is no exception. Note the rapid Ice Age transition 120,000 years ago, which is encircled in the graph of the North Greenland Ice Core Project.

Plasma density is the critical threshold factor

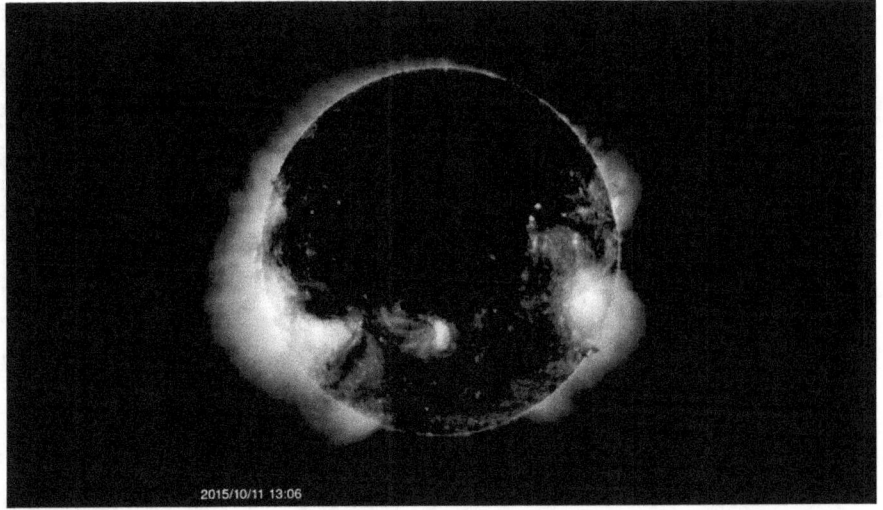

2015/10/11 13:06

For this phenomenon, the plasma density is the critical threshold factor, in the interstellar plasma stream that feed into the solar system.

We can study the fluctuations

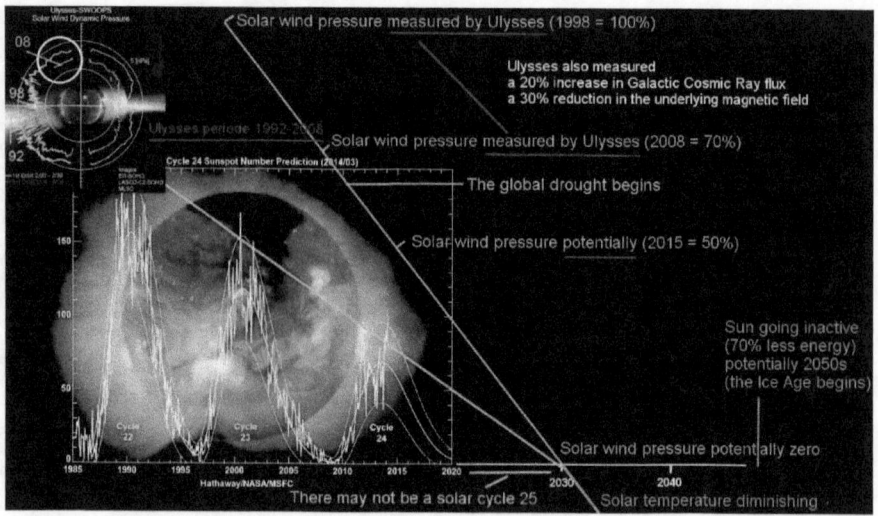

We can study the fluctuations and the trends and consequences, but not control them.

My studies in a number of video explorations

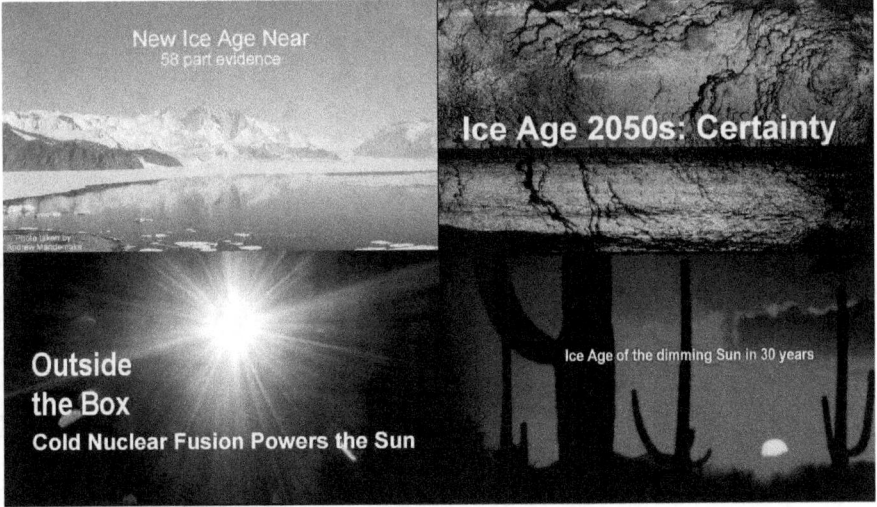

I have presented the relevant details of my studies in a number of video explorations. They all deal with the dynamics that unfold towards absolute climate change.

One of the items that are presented in the videos is important for the exploration here. It focuses critically on the coming Absolute Climate Change. This factor is important for determining how close we are in time to the Absolute Climate Change. This important factor is the solar wind.

The solar wind plays a 2-fold critical role

© Milloslav Druckmuller/Barcroft

http://www.zam.fme.vutbr.cz/~druck/Eclipse/ - an example of the amazing solar eclipse photography of Milloslav Druckmueller

The solar wind plays a 2-fold critical role in solar dynamics. It acts as a carrier that vents off plasma from the fusion cells. It also acts as a regulator when the positive feedback system delivers more plasma to the Sun than the plasma-fusion process requires. Thus, when the solar system runs strongly, strong solar winds result that carry off the excess plasma from the reaction cells. And as I said before, the solar winds also help in carrying the fusion products away from the Sun, so that its reaction cells won't clog up. Both functions are critical.

To explore the state of the solar wind

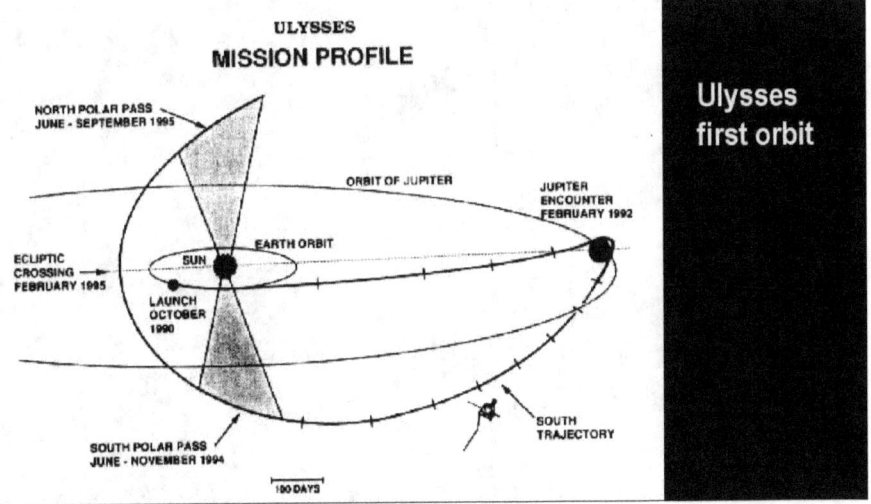

ULYSSES
MISSION PROFILE

Ulysses
first orbit

NORTH POLAR PASS
JUNE - SEPTEMBER 1995

ORBIT OF JUPITER

JUPITER
ENCOUNTER
FEBRUARY 1992

ECLIPTIC
CROSSING
FEBRUARY 1995

EARTH ORBIT

SUN

LAUNCH
OCTOBER
1990

SOUTH POLAR PASS
JUNE - NOVEMBER 1994

SOUTH
TRAJECTORY

100 DAYS

In order to explore the state of the solar wind as a critical component of the solar system, NASA had launched a spacecraft into a polar orbit around the Sun at a great distance from the Sun, as far as the distance to Jupiter. The spacecraft was named Ulysses. The mission planners had used the gravity of Jupiter to swing the satellite into a solar polar-trajectory. Here, something surprising happened when the satellite orbited over the polar regions. It encountered a void in the solar wind.

A steady solar wind at 800 km/second

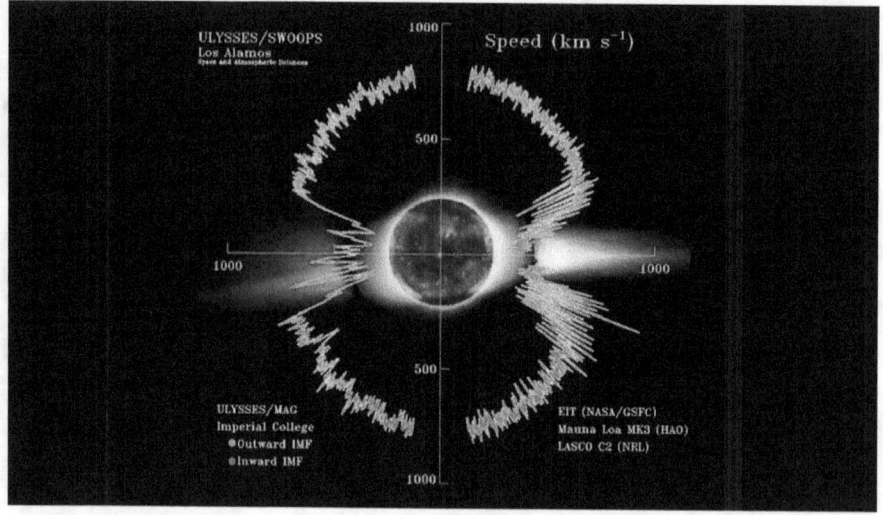

The satellite had measured a steady solar wind flowing away from the Sun at 800 km/second, except near the ecliptic where the solar wind gets messed up by the heliospheric current sheet that flows out from the Sun along the ecliptic, and in the polar regions were no solar wind was measured at all. This void is highly significant. It enables us to recognize a critical aspect of the plasma-powered Sun.

The void in solar wind is expected

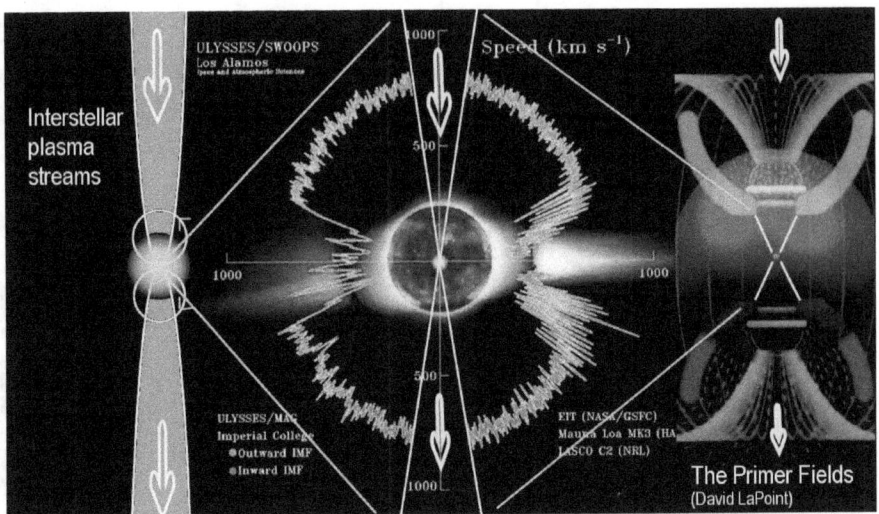

The void in solar wind is expected in the regions where the interstellar plasma stream interfaces with the plasma corona of the Sun, both for the inflowing and out flowing interface. As I said before, plasma streams become magnetically self-aligned as moving electric currents whereby they become increasingly concentrated by intense Primer Fields towards a sink point, from which in a weakened state they expand again and flow away through complimentary primer fields to become again an interstellar plasma streams flowing on to the next star.

The second orbit during a maximum in solar activity

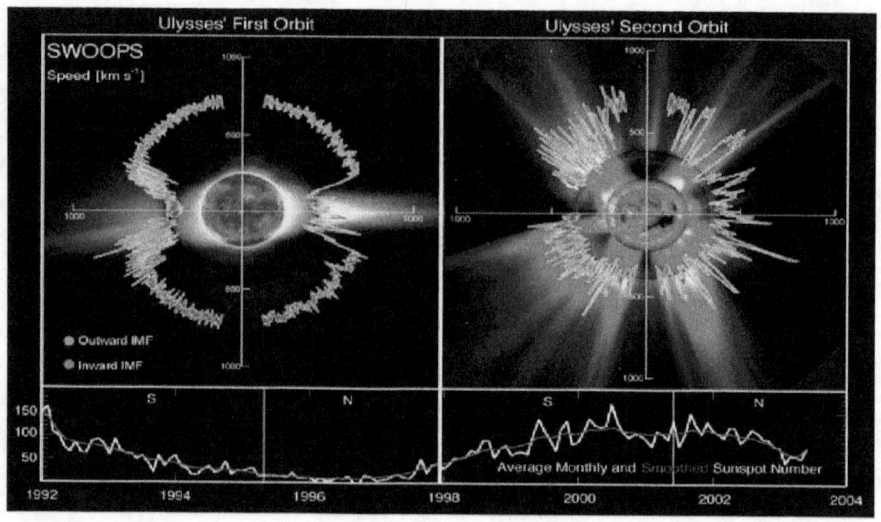

The solar wind that Ulysses had measured got messed up in a big way during most of the second orbit. The second orbit occurred during a maximum period in solar activity. The maximum period is typically a period of high plasma density in the solar corona that inhibits the solar wind, but which also fluctuates dramatically. Ulysses measurements of the solar wind-speed reflects these fluctuations.

The third surprise in the third orbit

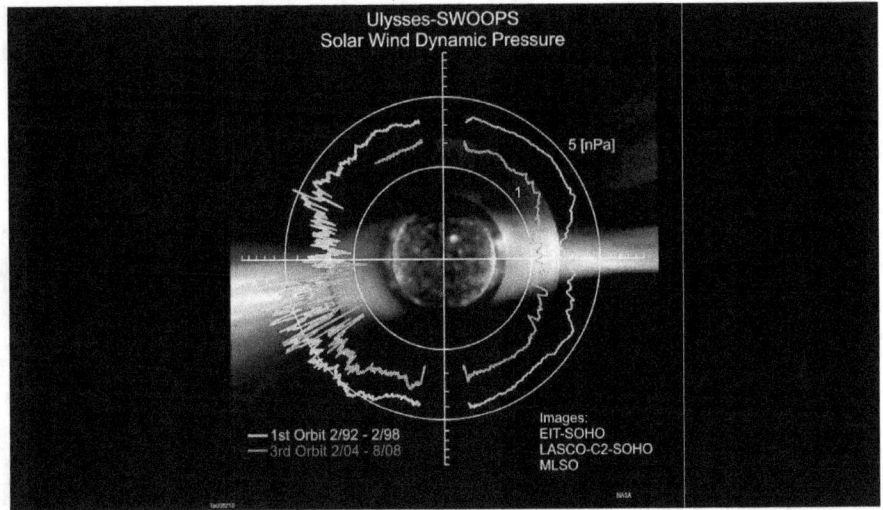

The third surprise was encountered in the third orbit, which again occurred during a quiet period in solar activity. The comparison between the 1st and 3rd orbit indicated that the solar wind-pressure was significantly less than it has been during the 1st orbit. The measured density and energy of the solar wind had diminished.

The solar wind pressure had diminished by 30%

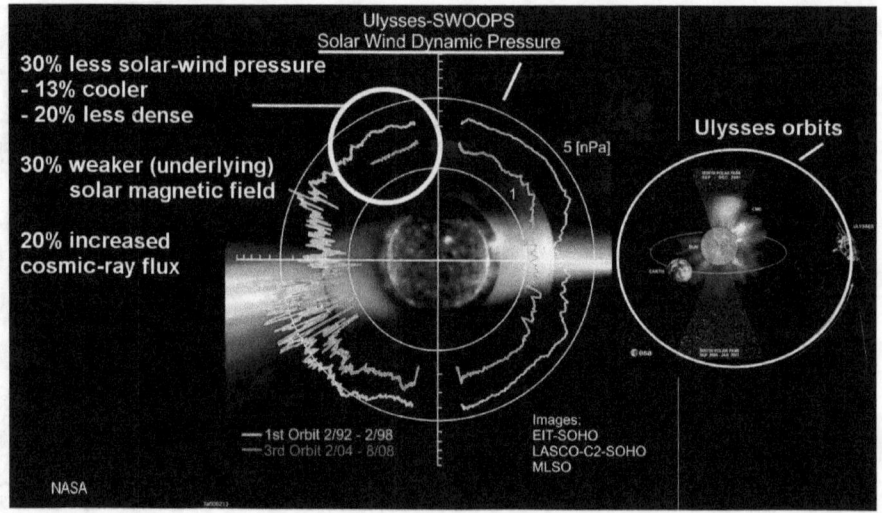

 The solar wind pressure had diminished by 30%, and the cosmic-ray flux had correspondingly increased by 20%. These are enormous changes for this short time span of just 12 years.

If one projects the diminishment forward

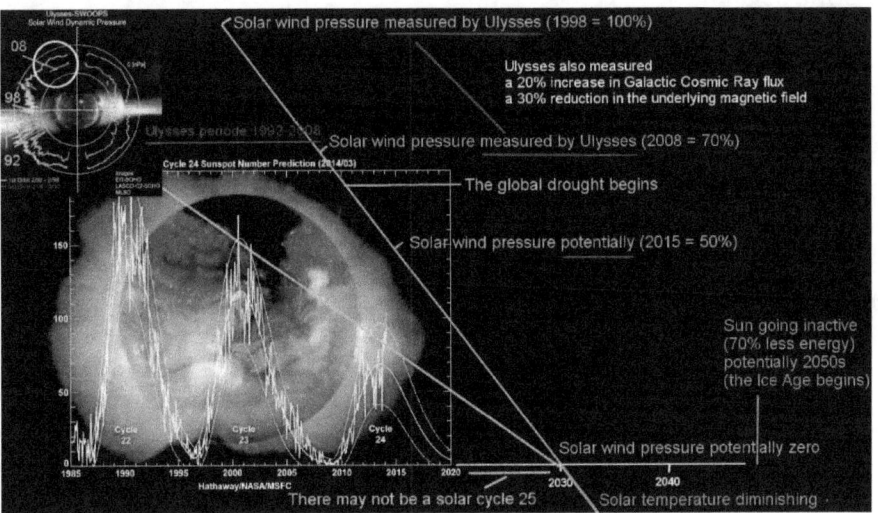

If one projects the measured rate of diminishment forward in a linear manner, the long-term projection would be that the solar wind would diminish to zero in the 2030s at the measured rate. That the diminishment is real, is also reflected in the observed rate of weakening of the solar cycles, Consequently, fear is being expressed now in some circles that another little ice age is coming up. However, the rapidly diminish solar wind pressure that Ulysses saw, suggests that this time around, possibly the entire positive feedback system that keeps the Sun's energy-level regulated within a narrow range of change, may shut down chaotically. It could potentially collapse before the interstellar plasma streams can recover.

The solar system might have come close to the point of collapse

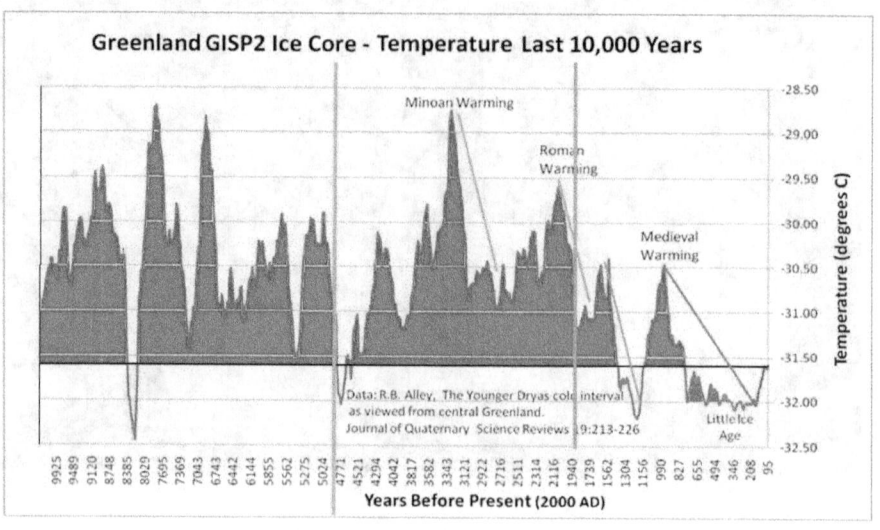

During the Little Ice Age in the 1600s, the solar system might have come close to the point of collapse, but was recovered before the chaotic collapse had occurred. The recovery was likely enabled by some substantial remnants in the interstellar plasma streams that had remained from the medieval warm period, the last major climate high-point. The background plasma from this period seems to have pulled the Earth out of its cosmic rut in the 1600s.

Now, we are facing the repeat of a major weakening process, but this time without a massive historic background backing us up. This means, with a high probability, that the climate fading that is now in progress may become an Absolute Change. The Sun may simply go inactive, probably without much of a warning. This is the characteristic of the collapse of positive feedback systems.

When the solar winds diminish to zero

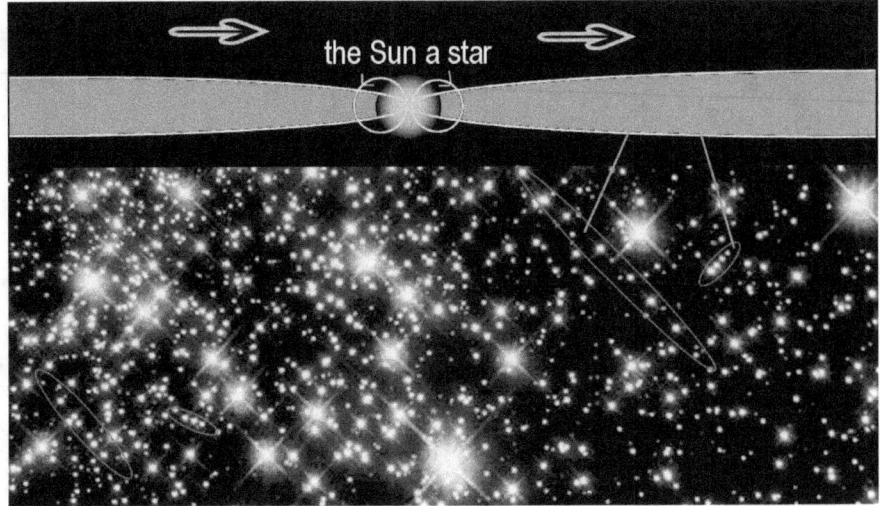

When the solar winds diminish to zero, we have reached the point when there is not enough plasma-density left in the interstellar plasma streams to keep the solar system regulated to its normal level. When this happens, when the interstellar plasma streams can't keep up with what the plasma fusion requires to keep the Sun operating, the Sun will diminish. Nobody knows what happens then. Without the solar winds flowing, which normally help to purge the fusion cells of their fusion products, large segments of fusion cells will likely clog up, collapse, and not recover.

What we see today in the form of coronal holes

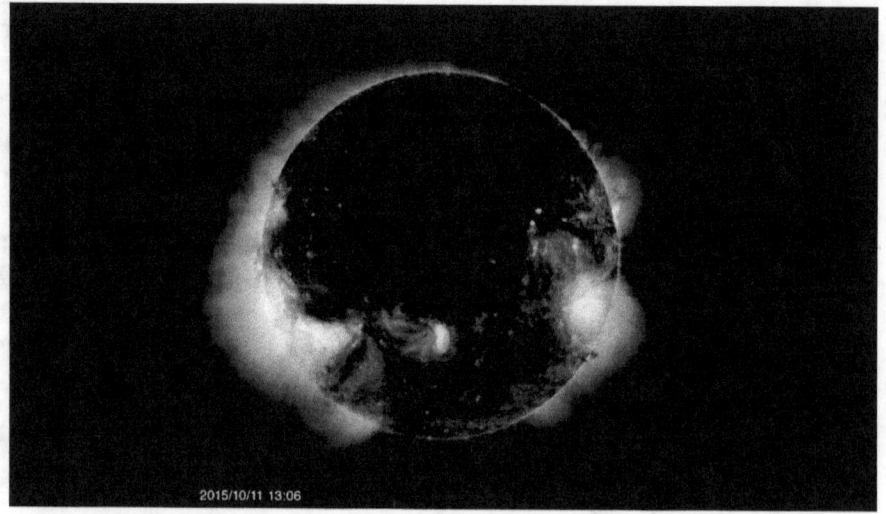

2015/10/11 13:06

What we see today in the form of coronal holes in the EUV band, may then the become the face of the Sun as a whole.

Plasma streams are electric streams

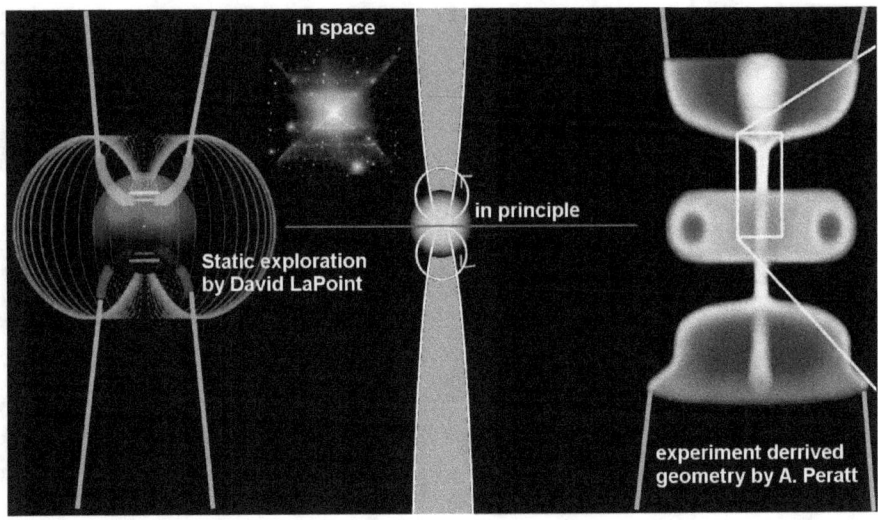

Interstellar plasma is focused around the Sun by magnetic phenomena. Magnetic phenomena are created by flowing electricity. Plasma streams are electric streams. The flowing plasma is consumed by the Sun in the process of synthesizing atomic elements. The resulting consumption of plasma maintains the flow that is needed for the magnetic fields to form, that focus the plasma unto the Sun. In this dynamic positive feed-back interplay, numerous parts of it support one-another. But, when one part fails, the entire dynamic system is affected and may stop functioning. When the magnetic fields diminish, plasma may no longer be focused onto the Sun. The plasma flow stops. The fields collapse. The Sun goes inactive.

When the Sun goes dim and cold

Ice Age of the dimming Sun in 30 years

www.ice-age-ahead-iaa.ca

When the Sun goes dim and cold, the next Ice Age begins. As I said before, the Sun will likely remain aglow for some time with residual energy within it. This may be the case until the plasma streams recover.

When we look through the umbra of the sunspots

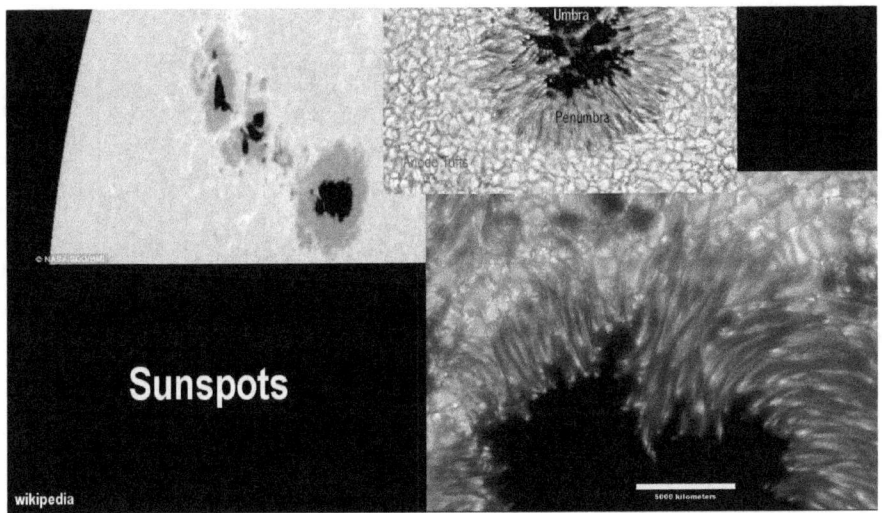

 When we look through the umbra of the sunspots, we see a 70% darker Sun inside. This default, darker face, may be the face of the Sun as a whole in the near future, beginning potentially in the 2050s timeframe or before.

It is certainly possible to survive at the 30% energy level

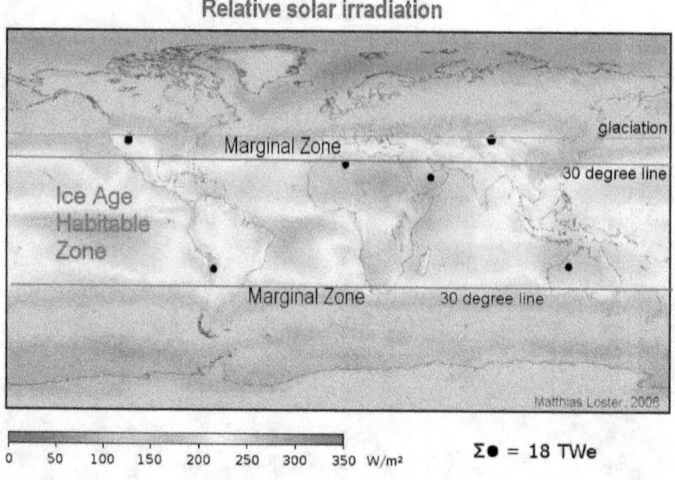

Relative solar irradiation

While it is certainly possible for society to survive at the 30% energy level, by relocating itself into the tropics where the solar irradiation is the greatest and is evidently sufficient for agriculture to be possible in a dimmer world, the question remains, will we stir our stumps and do it? Historically, humanity has survived the last Ice Age in the tropical regions, primarily in Africa, India, and Indonesia. Researchers begin to suggest that we may have survived there just barely.

We may have stood at the threshold to extinction

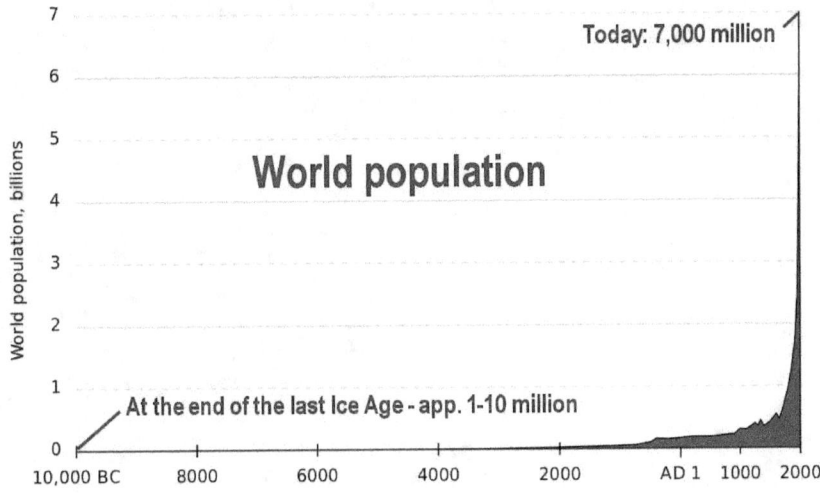

In general, researchers place the world population at the end of the last Ice Age into the range of 1-10 million people. That's all we had to show for, after 2 million years of human development. However, some researchers also suggest that the human population may have dwindled to much lower numbers at various stages during the harsh times of the deep glaciation, perhaps to the level of just a few thousand. We may have stood at the threshold to extinction many times during the ice ages. This is what we need to prevent this time around.

The progression towards the coming Ice Age

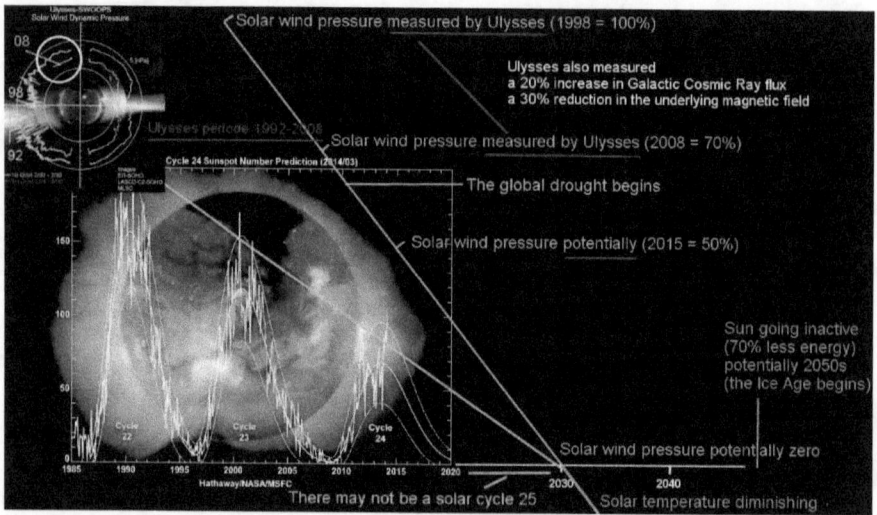

Sure, it is hard to imagine that we are presently in a transition zone to such 'dangerous' times, that may have been deadly times in the past, and may be deadly again if we fail to respond with appropriate preparations for them. Evidence suggests that the progression towards the coming Ice Age may be advancing more rapidly than we dare to acknowledge.

That the current weakening in the plasma environment

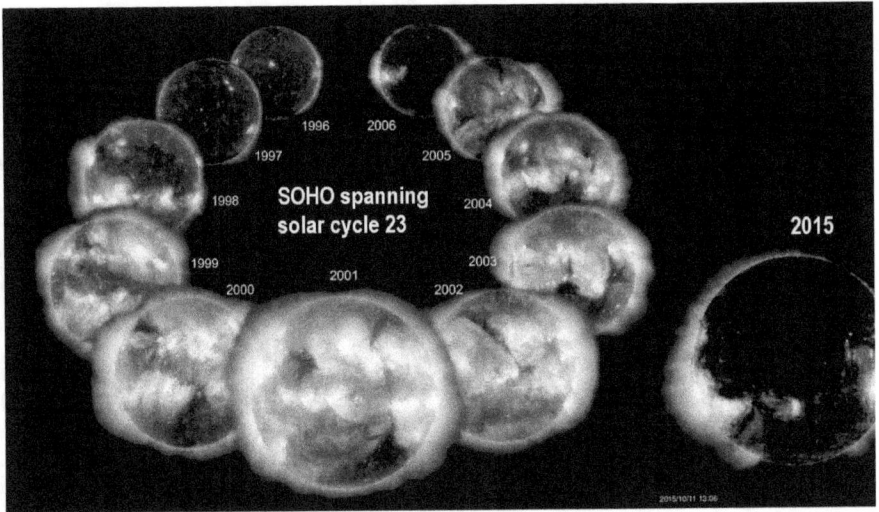

That the current weakening in the plasma environment towards an inactive Sun is dramatically progressing, becomes apparent when we compare a current coronal hole from 2015, on the lower right, with the 1997 image produced by the same satellite in the same color band. The 1997 image reflects the state of the Sun as it was at the lowest point of its respective solar cycle. Note that the 1997 image is substantially lighter than the coronal hole seen by the same camera in 2015, which, however, occurs at a point that is still 5 years distant from the weak point in the solar cycles.

We shouldn't see such massive coronal holes

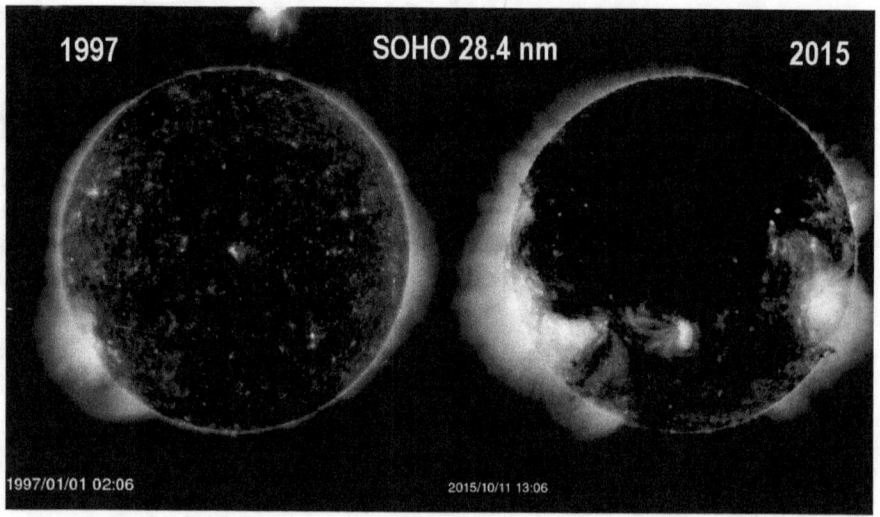

1997 SOHO 28.4 nm 2015

1997/01/01 02:06 2015/10/11 13:06

We shouldn't see such massive, deep-dark coronal holes that cover almost the entire Sun, and most definitely, we shouldn't see them so early in the solar cycle, 5 years from its low point. What we see here suggests that the plasma corona around the Sun is likely weakening more rapidly than the mere linear progression suggests, that has been chosen for the purpose of a comparative illustration.

While such infrastructures can be built

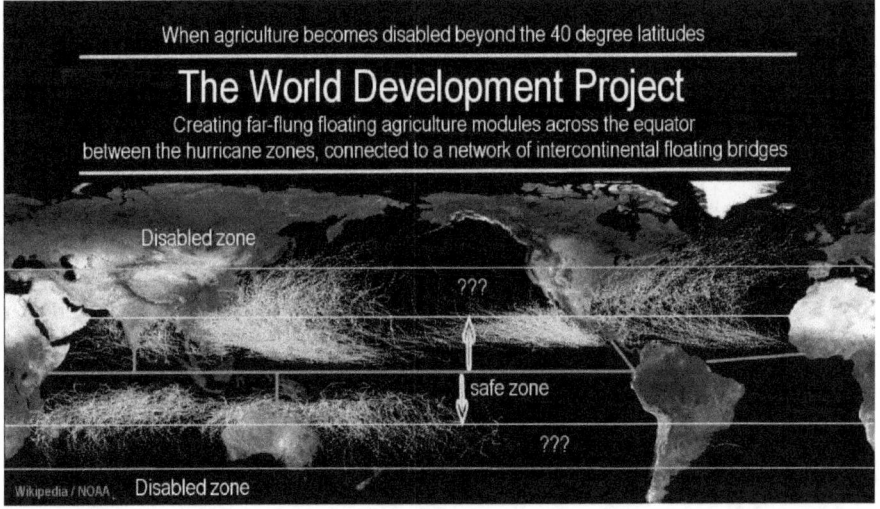

When agriculture becomes disabled beyond the 40 degree latitudes

The World Development Project

Creating far-flung floating agriculture modules across the equator
between the hurricane zones, connected to a network of intercontinental floating bridges

Disabled zone

???

safe zone

???

Wikipedia / NOAA   Disabled zone

In order to maintain our present 7 billion world population during the next Ice Age in possibly 30 years, massive infrastructures will have to be created in the tropics, much of it afloat on the seas, for the relocation of all the northern nations whose lands become uninhabitable in the cold darkness under an inactive Sun. While such infrastructures can be built, as are needed, and in time, created with automated industrial production for which the technologies, materials, and energy resources are all readily available, society may not find in its heart that human living is worthwhile enough to make the effort to assure its continued existence. That's what the current political and scientific landscape seems to suggest.

The Sun will not be affected by how society chooses

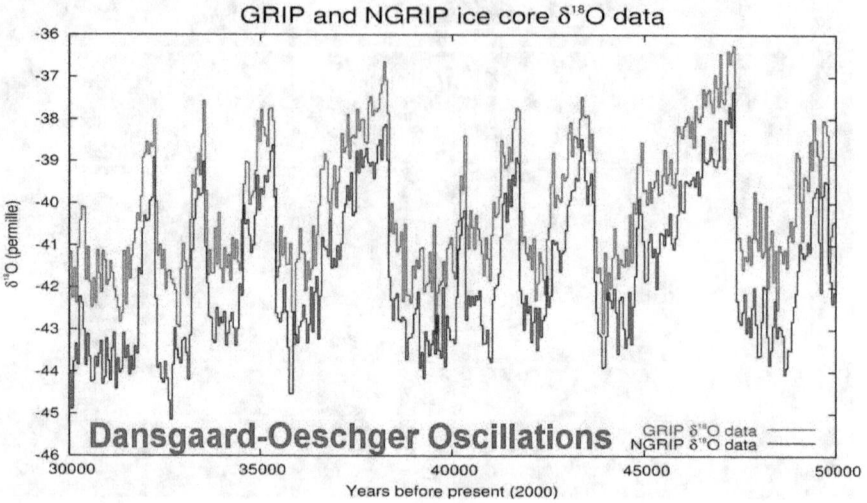

Whether or not the human world will make the effort to live, the Sun will not be affected by how society chooses. It will remain dormant for a season until the interstellar plasma streams recover for the Sun to restart. Such events seem to have occurred during the last Ice Age in 1470-year intervals. Once the Sun is re-lit, the available plasma become drawn down shortly thereafter, by which the Sun goes inactive again.

But what of it? We can live with that. For as long as we get 30% of the current solar radiation we can live easily in the tropics. The question remains, nevertheless, will we do it? Will we relocate the affected nations into the tropics? Or will humanity choose to let itself die?

The barrier that holds back the needed response

The barrier that holds back the needed response for our future in the near Ice Age is not so much a physical barrier than it is a mental barrier. And at the heart of that mental barrier stands still this diversity of the two fundamentally opposite perceptions of the nature of the Sun that I addressed at the beginning with the paradox of the sunlight. We face the same kind of paradox when we look into the future with the next Ice Age on the horizon in 30 years.

But will we solve the paradoxes?

Nothing exotic is needed on this front. Nothing more needs to be done than to implement in our heart and mind a revolutionary uplift in our self-perception as human beings. That's what the Ice Age Challenge impels us towards. Should we succeed on this front, the breakthrough promises to end wars, killing, and even the depopulation ideologies. All of this can be accomplished on the wings of an elevated sense of humanity and universal love.

I wrote a 12-part series of novels some years ago

The Lodging for the Rose

An exploration of
universal sovereignty
expressed in
universal love

A series of 12 novels
by Rolf A. F. Witzsche
free online, at:
ice-age-ahead-iaa.ca

I wrote a 12-part series of novels some years ago towards the greater realization of scientific honesty, primarily with ourselves. Scientific honesty in physics, will likely unfold in the course of greater honesty with ourselves as human beings. I named the series of novels "The Lodging for the Rose." I present the series for free in the hope that we, as a society of human beings, who are the pinnacle of life as we know it, will yet succeed in achieving what must be achieved for our survival, regardless of how unlikely this may presently seem. We have the capacity as human beings for revolutionary developments, even to create a brand new world of ourselves with such power on all fronts that the coming near Ice Age will have no effect on our living. We may be far from this at the present, especially in the sciences where our greatest challenge is to achieve a high level of scientific honesty. Still, I think we are on the course to be winning.

This simple image now stands poised to change the face of science

Just look at this image, a simple scene of common flowers of a type that everyone is familiar with. This simple image now stands poised to change the face of science, and with it the face of the world. With the truth poised to be unfolding, a new mental Climate for Absolute Change on all fronts is about to begin. We will no longer see the Sun isolated and disconnected, and consuming itself, and our humanity and civilization in a like manner. Instead we will see the Sun as a part of the ever-ongoing creative process of the universe, and we will see ourselves as a humanity in a lateral lattice of creative and productive human beings, joining hands in the building of new worlds, grander and brighter than anything we have yet dared to imagine.

Let me close this video with a verse of a poem from one of my novels. The title of the poem is; Harvest is Seedtime:

Love for one-another, the human spring

Love for one-another, the human spring
Mankind is afloat in a sea that is Love
Seeds germinate, become plants
Roots break the ground
Love lifts the barriers, patiently
Silently waiting, reaching for the sky

Love for one-another, the human spring,
Mankind is afloat in a sea that is Love,
Seeds germinate, become plants,
Roots break the ground,
Love lifts the barriers, patiently,
Silently waiting, reaching for the sky.